LECTURE NOTES
in
NUMERICAL
ANALYSIS
with

MATHEMATICA

Krystyna STYŠ & Tadeusz STYŠ

Contents

0.1 FOREWORD

The lecture notes have been developed during over thirty years of teaching experience by the authors in the area of Numerical Analysis. They have taught the course on Numerical Methods designed for Science and Engineering students, at the University of Warsaw, University of Jos, Nigeria and the University of Botswana. The content of the notes covers the following topics: Computer Numbers and Round-off Errors Estimates, Interpolation and Approximation of Functions, Polynomial Splines and Applications, Numerical Integration and Solution of Non-linear Equations. The authors have presented the subjects in exact and comprehensive way with emphasis put on formulation of fundamental theorems with proofs supported by well selected examples. They used Mathematica, System for doing Mathematics, in solving problems specific to the subjects. In the notes, the reader will find interesting algorithms and their implementation in the Mathematica System. The lecture notes are well written and recommended as reading material for a basic course in Numerical Methods for science and engineering students.

O. A. Daman
University of Botswana,
Botswana

0.2   PREFACE

This text is intended for science and engineering students. It covers most of the topics taught in a first course on numerical analysis and requires some basic knowledge in calculus, linear algebra and computing. The text has been used as recommended handbook for courses taught on numerical analysis at undergraduate level. Each chapter ends with a number of questions. It is taken for granted if the reader has access to computer facilities for solving some of these questions using *Mathematica*. There is extensive literature published on numerical analysis including books on undergraduate and postgraduate levels. As the text for a first course in numerical analysis, this handbook contains classical content of the subject with emphases put on error analysis, optimal algorithms and their implementation in computer arithmetic. There is also a desire that the reader will find interesting theorems with proofs and verity of examples with programs in *Mathematica* which help reading the text. The first chapter is designed for floating point computer arithmetic and round-off error analysis of simple algorithms. It also includes the notion of well conditioned problems and concept of complexity and stability of algorithms.

Within chapter 2, interpolation of functions is discussed. The problem of interpolation first is stated in its simplest form for polynomials, and then is extended to generalized polynomials. Different Chebyshev's systems for generalized interpolating polynomials are considered.

In chapter 3, polynomial splines are considered for uniform approximation of an one variable function.

Fundamental theorems on uniform approximation (Taylor's theorem, Weierstrass theorem, Equi-Oscillation Chebeshev's theorem) are stated and proved in chapter 4.

Chapter 5 is an introduction to the least squares method and contains approximation of functions in the norm of $L_2(a, b)$ space. Also, it contains approximation of discrete data and an analysis of experimental data.

In the chapter 6, two techniques of numerical integration are developed, the Newton-Cotes methods and Gauss methods. In both methods an error analysis is carried out.

For solution of non-linear algebraic equations, the most popular methods, such as Fixed Point Iterations, Newton Method, Secant Method and Bisection Method, are described in chapter 7.

ACKNOWLEDGEMETNTS
Declared None
CONFLICT OF INTEREST
The authors confirm that this chapter contents have no conflict of interest

Krystyna STYŠ
University of Warsaw, Poland
Tadeusz STYŠ
University of Warsaw, Poland

The List of Mathematica Functions and Modulae

LECTURE NOTES
in
NUMERICAL
ANALYSIS
with

MATHEMATICA

Send Orders for Reprints to reprints@benthamscience.net

CHAPTER 1

Floating Point Computer Arythmetic

Abstract

In this chapter, propagation of round-off errors in floating point arithmetic operations of computers numbers is presented. The notions of the conditional number, stability and complexity of algorithms are introduced and illustrated by examples. The Horners scheme for evaluation of Polynomials is given to elucidate the optimal and well-conditioned algorithms when they are implemented in a computer system like Mathematica. The chapter ends with a set of questions.

Keywords: Computer numbers, Round-off errors.

1.1 Computer Representation of Numbers

Normally, computers operate in floating point arithmetic, and decimal numbers are presented in the following form:

$$x = \mp m 10^c,$$

Krystyna STYŠ & Tadeusz STYŠ

where $m = 0.\alpha_1\alpha_2\ldots\alpha_r$; $\alpha_1 \neq 0$; $0 \leq \alpha_i \leq 9$; $i = 1, 2, \ldots, r$ is called mantissa of x, and the integer c is called exponent of x. The most significant digit α_1 is always different from zero. Therefore, mantissa m satisfies the following inequality:

$$0.1 \leq m < 1.$$

Clearly, a number can have exact floating point computer representation if its mantissa consists of a finite number of digits. For example $1/4$ has exact floating point computer representation and its mantissa $m = 0.25$, while the power $c = 0$. However

$$\frac{1}{3} = 0.333\ldots$$

cannot be accepted as a floating point number by a computer since its mantissa $m = 0.333...$, has infinite number of digits.

Nevertheless, any real number, even with infinite number of digits, can be also represented in a computer floating point arithmetic with certain round-off error $\epsilon \leq \frac{1}{2} 10^{-r}$. For example

$$x = \frac{2}{3} = 0.66666666666...$$

has the 4-digit floating point representation

$$fl(x) = 0.6667$$

and its round-off error $\epsilon = 0.0000333...$

In order to round off in **Mathematica** at fourth decimal place, we execute the following instruction (*cf.* [28])

$$\texttt{N[Round[}\frac{2}{3}\texttt{10}^4\texttt{]/10}^4\texttt{]} = 0.6667.$$

[1] Below, we shall consider round-off up operation of a floating point number (cf. [5, 15])

$$x = \pm m \ 10^c$$

---

[1]Note that in computers round off down rule is used

at the r-th decimal place.

In the round-off up operation x is replaced by $fl(x)$, so that

$$fl(x) = \pm \overline{m} \ 10^{\overline{c}},$$

where the new mantissa

$$\overline{m} = fl(m) = \begin{cases} [m * 10^r + 0.5] * 10^{-r} & if \ \ \overline{c} = c, \\ [m * 10^r + 0.5] * 10^{-r+1} & if \ \ \overline{c} = c+1. \end{cases}$$

[2] Clearly, the exponent $\overline{c} = c + 1$ if $\alpha_1 = \alpha_2 = \cdots = \alpha_r = 9$ and $\alpha_{r+1} \geq 5$, otherwise c does not change its value.

For example, if $x = 0.99999 * 10^{-2}$ and $r = 4$, then the mantissa

$$\overline{m} = fl(0.99999) = 0.1000,$$

and the exponent

$$\overline{c} = c + 1 = -2 + 1 = -1.$$

Thus, the 4 digit floating point representation of x is

$$fl(0.99999 * 10^{-2}) = 0.1000 * 10^{-1}.$$

Also, for $x = 0.234584 * 10^2$ and $r = 3$, we have

$$\overline{m} = [0.234584 * 10^3 + 0.5] * 10^{-3} = 0.235,$$

and

$$\overline{c} = c = 2.$$

Hence, the 3 digit floating point representation of x is

$$fl(0.234584 * 10^2) = 0.235 * 10^2.$$

The Absolute and Relative Round-off Errors *The absolute round-off error of a real number* $x = \mp m 10^c$ is

$$\epsilon_x = fl(x) - x$$

---

[2] Here $[y]$ is the greatest integer not greater than y.

This error satisfies the inequality

$$| \, fl(x) - x \, | \leq \epsilon * 10^c,$$

where $\epsilon = \frac{1}{2} 10^{-r}$. For example, let $x = 0.57367864 * 10^2$ and $r = 3$. Then, we have

$$| \, fl(0.57367864 * 10^2) - 0.57367864 * 10^2 \, | =$$

$$| \, 0.574*10^2 - 0.57367864*10^2 \, | = 0.032136 < \frac{1}{2} 10^{-3}*10^2 = 0.05.$$

The relative round-off error of a floating point number $x = \mp m \, 10^c \neq 0$ is

$$\delta_x = \frac{fl(x) - x}{x} \quad \text{if} \ \ x \neq 0.$$

Because $m \geq 0.1$, therefore

$$\left| \frac{| \, fl(x) - x \, |}{x} \right| \leq \frac{1}{2} 10^{1-r}, \quad x \neq 0.$$

Indeed, we have

$$\left| \frac{fl(x) - x}{x} \right| = \left| \frac{fl(\mp m 10^c) \pm m 10^c}{\mp m 10^c} \right| \leq 10\epsilon = \frac{1}{2} 10^{1-r}.$$

Thus, the relative round-off error does not exceed $\delta = \frac{1}{2} 10^{1-r}$ which is called *computer precision*.

For example, if $r = 3$ then the computer precision $\delta = \frac{1}{2} 10^{-2}$ and the relative round-off error of $x = 0.57367864 * 10^2$ is

$$\left| \frac{fl(x) - x}{x} \right| = \frac{0.032136}{0.57367864 * 10^2} = 0.0005601742.$$

The relative round-off error is closely related to the *percent round-off error*,

$$p\% = 100 * \delta_x\% = 100 \frac{fl(x) - x}{x}\% \quad \text{if} \ \ x \neq 0.$$

Thus, the percent round-off error of the number x in the above example is

$$p\% = 100 * 0.5601742 * 10^{-3}\% = 0.5601742 * 10^{-1}\%.$$

Obviously, the results of arithmetic operations $x \pm y$, xy and x/y might not be computer numbers, even if x and y are given in the r-digit floating point computer arithmetic. For example, $x = 0.11111111$ and $y = 0.55555555$ are 8-digit computer numbers, (8-digit mantissa). But the product $xy = 0.617283938271605 * 10^{-1}$ has 15-digit mantissa which must be rounded-off up to 8 digits to be accepted by an 8-digit computer.

Let us assume that x and y are r-digit floating point numbers. Then, the relative round-off errors of the four arithmetic operations $+$, $-$, $*$, $/$ are:

$$\epsilon = \frac{fl(x \pm y) - (x \pm y)}{x \pm y} \qquad if \quad x \pm y \neq 0,$$

$$\eta = \frac{fl(xy) - xy}{xy} \qquad if \quad xy \neq 0,$$

$$\gamma = \frac{fl(x/y) - x/y}{x/y} \qquad if \quad xy \neq 0.$$

Hence

$$fl(x \pm y) = (x \pm y)(1 + \epsilon),$$

$$fl(xy) = xy(1 + \eta),$$

$$fl(x/y) = x/y(1 + \gamma) \qquad if \quad y \neq 0,$$

where the absolute values of ϵ, η and γ do not exceed computer precision $\delta = \frac{1}{2}10^{1-r} = 5 * 10^{-r}$.

Now, let us assume that x and y are not computer numbers and they have the following floating point representation:

$$\overline{x} = fl(x) \qquad and \quad \overline{y} = fl(y).$$

Then, the round-off errors

$$|\Delta x| = |x - \overline{x}| \leq \beta \quad \text{and} \quad |\Delta y| = |y - \overline{y}| \leq \beta$$

do not exceed certain $\beta > 0$.

Below, we shall estimate propagation of these errors in the four arithmetic operations.

For addition and subtraction

$$|(x \pm y) - (\overline{x} \pm \overline{y})| = |\Delta x \pm \Delta y| \leq 2\beta.$$

For multiplication

$$|xy - \overline{xy}| = |(\overline{x} + \Delta x)(\overline{y} + \Delta y) - \overline{xy}|.$$

Hence, for small Δx and Δy the term $\Delta x \Delta y \approx 0$, and then the error estimate is:

$$|xy - \overline{xy}| \leq |\overline{x}| \, |\Delta y| + |\overline{y}| \, |\Delta x| \leq (|\overline{x}| + |\overline{y}|)\beta.$$

For division, when $y \neq 0$, we have

$$
\begin{aligned}
|\frac{x}{y} - \frac{\overline{x}}{\overline{y}}| &= |\frac{\overline{x} + \Delta x}{\overline{y} + \Delta y} - \frac{\overline{x}}{\overline{y}}| \\
&= |\frac{\overline{xy} + \overline{y}\Delta x - \overline{xy} - \overline{x}\Delta y}{\overline{y}(\overline{y} + \Delta y)}| \\
&\leq \frac{|\overline{x}| \, |\Delta y| + |\overline{y}| \, |\Delta x|}{|\overline{y}|^2(1 - \frac{\Delta y}{\overline{y}})}.
\end{aligned}
$$

Hence, for small $\dfrac{\Delta y}{\overline{y}} \approx 0$, we arrive at the following error estimate in division

$$|\frac{x}{y} - \frac{\overline{x}}{\overline{y}}| \leq \frac{|\overline{x}| \, |\Delta y| + |\overline{y}| \, |\Delta x|}{\overline{y}^2} \leq \frac{|\overline{x}| + |\overline{y}|}{\overline{y}^2}\beta.$$

1.2   Conditional Number

Conditional number is naturally affiliated with a problem under consideration. This number is used to assess perturbation of the final results caused by round-off errors or errors in measurement of input data. For example, the problem of evaluating of a differentiable function $f(x)$ for a given x leads to the conditional number determined below. Namely, if x^* is an approximate value to x then the relative errors

$$\frac{f(x) - f(x^*)}{f(x)} \quad \text{and} \quad \frac{x - x^*}{x}$$

determine the *conditional number*

$$Cond(f(x)) = \lim_{x^* \to x} \frac{\frac{f(x) - f(x^*)}{f(x)}}{\frac{x - x^*}{x}} = x \frac{f'(x)}{f(x)}, \quad x \neq 0, \quad f(x) \neq 0,$$

at a point x.

Thus, the conditional number is expressed in terms of $f(x)$ and $f'(x)$ and it is independent of an algorithm used to evaluate $f(x)$.

Example 1.1 *Let $f(x) = x^\alpha$, for a real α. We get*

$$Cond(f(x)) = x \frac{f'(x)}{f(x)} = x \frac{\alpha x^{\alpha-1}}{x^\alpha} = \alpha.$$

Hence, the conditional number of the power function equals to the constant α.

Let us note the following relations which hold for two differentiable functions:

$$Cond(f(x)g(x)) = Cond(f(x)) + Cond(g(x)),$$

$$Cond(\frac{f(x)}{g(x)}) = Cond(f(x)) - Cond(g(x)).$$

Indeed, we have

$$
\begin{aligned}
Cond(f(x)g(x)) &= x \frac{f'(x)g(x) + f(x)g'(x)}{f(x)g(x)} \\
&= Cond(f(x)) + Cond(g(x)),
\end{aligned}
$$

and

$$Cond(\tfrac{f(x)}{g(x)}) = x\frac{f'(x)g(x) - f(x)g'(x)}{g^2(x)}\frac{g(x)}{f(x)}$$
$$= Cond(f(x)) - Cond(g(x)).$$

A problem is said to be *well-conditioned* if its conditional number is relatively small, otherwise it is referred to as *ill-conditioned* problem.

In order to solve a problem, we should choose an algorithm which is stable, that is, resistant against round-off errors accumulation. Here, we consider stability with respect to round-of errors accumulation, in the sense of the following definition:

Definition 1.1 *An algorithm is said to be stable with respect to round-off errors accumulation, if it produces a solution biased by error proportional to a computer precision.*

Normally, stable algorithms are known for well-conditioned problems, that is, for problems with relatively small conditional numbers.

Example 1.2 *Let us evaluate*

$$f(x) = x^{1024},$$

at $x = 0.999$, in three digits floating point arithmetic, by the following two algorithms:

Algorithm I .

- *Let $x := 1$.*

- *For i:=1 to 1024,*
  compute $x := 0.999 * x$,
  in three digit floating point arithmetic.

It is clear that the conditional number of the product is equal to 1024. The result obtained by the algorithm is equal to 0.5.

Algorithm II .

- *Let $x := 0.999, \quad i := 0$;*

- *Repeat*

$$x := x * x,$$

$$i := i + 1,$$

until $i = 10$;

The conditional number of the product is 20. The result of the second algorithm is equal to 0.3530

Comparing both the results with the exact value $f(0.999) = 0.3590$, we note that the result from the first algorithm is less accurate than the one obtained by the second algorithm. This is due to relatively small conditional number involved in the second algorithm.

A part of stability, an algorithm is also characterized by the number of operations needed to transform input data into expected results. This number of operations is called *computational complexity* of an algorithm. In the above example, the computational complexity of the first algorithm is equal to 1024 operations, and the computational complexity of the second algorithm is equal to 10 operations.

1.3   Round-off Error of the Sum $\sum x_i$.

Let $x_i = \mp m_i 10^{c_i}, \quad i = 1, 2, \ldots, n$; be r-digit floating point numbers and let ϵ_i be round-off errors associated with adding x_i to the previously evaluated sum $x_1 + x_2 + \cdots + x_{i-1}$ which do not exceed the computer precision $\delta = \frac{1}{2} 10^{1-r}$, *i.e.*

$| \epsilon_i |, \leq \delta$, $i = 1, 2, \ldots, n$. Then, we have

$$fl(x_1 + x_2 + \cdots + x_n) =$$

$$= (\cdots (x_1 + x_2)(1 + \epsilon_2) + x_3)(1 + \epsilon_3) + \cdots + x_n)(1 + \epsilon_n)$$

$$= \sum_{i=1}^{n} x_i \prod_{j=i}^{n} (1 + \epsilon_j), \text{ when } \epsilon_1 = 0,$$

and

$$|fl(\sum_{i=1}^{n} x_i) - \sum_{i=1}^{n} x_i| = |\sum_{i=1}^{n} x_i [\prod_{j=i}^{n} (1 + \epsilon_j) - 1]|$$

$$\leq \sum_{i=1}^{n} | x_i |[(1 + \delta)^{n-i+1} - 1]$$

$$\leq [(1 + \delta)^n - 1] \sum_{i=1}^{n} |x_i|.$$

Since, for small δ, we have

$$(1 + \delta)^n - 1) \approx n\delta,$$

the absolute and relative round-off errors of the sum $x_1 + x_2 + \ldots + x_n$ satisfy the following inequalities:

$$|fl(\sum_{i=1}^{n} x_i) - \sum_{i=1}^{n} x_i| \leq n\delta \sum_{i-1}^{n} |x_i|,$$

$$\frac{|fl(\sum_{i=1}^{n} x_i) - \sum_{i=1}^{n} x_i|}{| \sum_{i=1}^{n} x_i|} \leq n\delta \frac{\sum_{i=1}^{n} |x_i|}{| \sum_{i=1}^{n} x_i|}.$$

For example, let us compute the sum

$$98.7 + 2.48 + 0.875$$

in 3-digit floating point arithmetic

$$fl(98.7 + 2.48 + 0.875) = fl(fl(98.7 + 2.48) + 0.875$$

$$= fl(101 + 0.875) = 102.$$

The exact result

$$98.7 + 2.48 + 0.875 = 102.055.$$

and the absolute round-off error is equal to 0.055
The relative round off error

$$\frac{fl(98.7 + 2.48 + 0.875) - (98.7 + 2.48 + 0.875)}{102.055} = \frac{0.055}{102.055}$$
$$= 0.0005363.$$

Example 1.3 *Consider the following function:*

$$g(x, y, z) = x + y + z, \qquad x, y, z \in (0, \infty).$$

(i). *Evaluate* $g(0.95, 0.475, 0.625)$ *in 3-digit floating point arithmetic. Determine absolute and relative round-off errors.*

(ii). *Give an estimate of absolute and relative round-off errors for evaluation of the function* $g(x, y, z)$ *assuming that* $x, y, z \in (0, \infty)$ *are 3-digit floating point numbers.*

Solution (i). We compute

$$fl(0.95 + 0.475 + 0.625) = fl(fl(0.95 + 0.475) + 0.625)$$
$$= fl(0.143 * 10 + 0.625)$$
$$= 0.206 * 10.$$

The exact value of the sum

$$0.95 + 0.475 + 0.625 = 2.05.$$

Thus, the absolute error

$$| fl(0.95 + 0.475 + 0.625) - (0.95 + 0.475 + 0.625) |$$
$$= 2.06 - 2.05 = 0.01,$$

and the relative round-off error

$$\frac{fl(0.95 + 0.475 + 0.625) - (0.95 + 0.475 + 0.625)}{0.95 + 0.475 + 0.625}$$

$$= \frac{0.01}{2.05} = 0.004878.$$

Using `Mathematica`, when round off down rule is used, the result is

$$N[0.95 + 0.475 + 0.625, 3] = 2.05.$$

Solution (ii). The computer precision

$$\delta = \frac{1}{2}10^{1-3} = 0.005.$$

Let ϵ_2 and ϵ_3 be round-off errors which may appear when the two additions are executing *i.e.* $\mid \epsilon_2 \mid \leq \delta$ and $\mid \epsilon_3 \mid \leq \delta$. Then, we have

$$fl(x + y + z) \ = [(x + y)(1 + \epsilon_2) + z](1 + \epsilon_3)$$

$$= (x + y)(1 + \epsilon_2)(1 + \epsilon_3) + z(1 + \epsilon_3)$$

$$= x + y + z + (x + y)(\epsilon_2 + \epsilon_3 + \epsilon_2\epsilon_3) + z\epsilon_3.$$

Since $\epsilon_2\epsilon_3$ is small compared with ϵ_2 or ϵ_3 then the term $\epsilon_2\epsilon_3 \approx 0$ may be ignored. Therefore, the absolute round-off error estimate is

$$\mid fl((x + y + z) - (x + y + z)) \mid \ \leq \ 2\delta(x + y) + \delta z$$

$$\leq \ 2\delta(x + y + z)$$

$$\leq \ 0.01\,(x + y + z),$$

for all 3-digit floating point numbers $x, y, z \in (0, \infty)$. Hence, the estimate of relative round-off error of the sum is

$$\mid \frac{fl(x + y + z) - (x + y + z)}{x + y + z} \mid \leq \frac{0.01\,(x + y + z)}{x + y + z} = 0.01,$$

for all 3-digit floating point numbers $x, y, z \in (0, \infty)$, $x + y + z \neq 0$.

1.4   Round-off Error of the Product $\prod x_i$.

As in the sum $\sum x_i$, we assume that $x_i = \mp m_i \, 10^{c_i}$, $i = 1, 2, \ldots, n$; are r-digit floating point numbers and the relative round-off errors η_i in the multiplications do not exceed the computer precision $\delta = \frac{1}{2} 10^{1-r}$, $i.e.$ $\mid \eta_i \mid \leq \delta$; $i = 1, 2, \ldots, n$. We then arrive at

$$fl(x_1 x_2 \cdots x_n) = x_1 x_2 \cdots x_n (1 + \eta_2) \cdots (1 + \eta_n)$$

$$= \prod_{i=1}^{n} x_i \prod_{k=1}^{n} (1 + \eta_k) \quad \text{when} \quad \eta_1 = 0.$$

Hence, we obtain the following estimate of the absolute round-off error of the product

$$\mid fl(\prod_{i=1}^{n} x_i) - \prod_{i=1}^{n} x_i \mid = \mid [\prod_{i=1}^{n}(1 + \eta_i) - 1] \prod_{i=1}^{n} x_i \mid$$

$$\leq ((1 + \delta)^n - 1) \mid \prod_{i=1}^{n} x_i \mid \leq n\delta \mid \prod_{i=1}^{n} x_i \mid .$$

Therefore, the relative round-off error satisfies the following inequality:

$$\frac{\mid fl(\prod_{i=1}^{n} x_i) - \prod_{i=1}^{n} x_i \mid}{\mid \prod_{i=1}^{n} x_i \mid} = \mid \prod_{i=1}^{n}(1 + \eta_i) - 1 \mid$$

$$\leq (1 + \delta)^n - 1 \leq n\delta.$$

For example, we may evaluate the product

$$57.6 * 4.68 * 0.385$$

in 3-digit floating point arithmetic as follows:

$$fl(57.6 * 4.68 * 0.385) = fl(fl(57.6 * 4.68) * 0.385)$$

$$= fl(270 * 0.385) = 104.$$

The exact result is

$$57.6 * 4.68 * 0.385 = 103.78368.$$

The absolute round-off error

$$| \, fl(57.6 * 4.68 * 0.385) - (57.6 * 4.68 * 0.385) \, |$$

$$= 104 - 103.78368 = 0.21632$$

$$\left| \frac{fl(57.6 * 4.68 * 0.385) - (57.6 * 4.68 * 0.385)}{57.6 * 4.68 * 0.385} \right|$$

$$= \frac{0.21632}{103.78368} = 0.00208.$$

Example 1.4 *Consider the following function:*

$$g(x, y, z) = x \, y \, z, \quad \text{for} \quad x, y, z \in (0, \infty).$$

(i). *Evaluate $g(3.75, 5.86, 8.65)$ in 3-digit floating point arithmetic. Determine absolute and relative round-off errors.*
(ii). *Give an estimate of absolute and relative round-off errors assuming that $x, y, z \in (0, \infty)$ are computer numbers and the computer precision $\delta = 0.005$.*

Solution (i). We compute

$$fl(3.75 * 5.86 * 8.65) = fl((3.75 * 5.86) * 8.65))$$

$$= fl(22 * 8.65) = 190.$$

The exact result

$$3.75 * 5.86 * 8.65 = 190.08375.$$

The absolute round-off error

$$| \, fl(3.75 * 5.86 * 8.65) - 3.75 * 5.86 * 8.65 \, |$$

$$= | \, 190 - 190.08375 \, | = 0.08375.$$

and the relative round-off error

$$\left| \frac{fl(3.75 * 5.86 * 8.65) - 3.75 * 5.86 * 8.65}{3.75 * 5.86 * 8.65} \right|$$

$$= \frac{0.08375}{190.08375} = 0.000440595.$$

In **Mathematica**, we obtain the same result when the following statement is executed

```
N[3.75*5.86*8.65,3];
```

Solution (ii). In order to estimate the absolute and relative round-off errors for arbitrary $x, y, z \in (-\infty, \infty)$, we find

$$\begin{aligned}|fl(x*y*z) - x*y*z| &= |\ x*y*(1+\eta_2)*z*(1+\eta_3) \\ &\quad - x*y*z\ | \\ &= |x*y*z*(\eta_2 + \eta_3 + \eta_2*\eta_3)|.\end{aligned}$$

Since $\eta_2 * \eta_3 \approx 0$, the absolute round-off error estimate is

$$|\ fl(x*y*z) - x*y*z\ | \le 2\delta|x*y*z| = 0.01*|x*y*z|$$

and relative round-off error estimate is

$$\left|\ \frac{fl(x*y*z) - x*y*z}{x*y*z}\ \right| \le 2\delta = 0.01.$$

for all $x, y, z \in (-\infty, \infty)$ and $xyz \ne 0$.

1.5 Horner Scheme

In order to compute the value of the polynomial

$$P_n(x) = a_n x^n + a_{n-1} x^{n-1} + \ldots + a_1 x + a_0$$

at a point x, we can use the Horner's scheme,

$$P_n(x) = (\ldots((a_n x + a_{n-1})x + a_{n-2})x + \ldots + a_1)x + a_0$$

which leads to the following algorithm:

Let
$$p = a_n,$$
　then
$$for \ \ i = n - 1, n - 2, \ldots, 0,$$
　evaluate
$$p := px + a_i.$$

As a result of execution of this algorithm, we obtain the value of $P_n(x)$ for n multiplications and n additions. Later on, we shall show that the total number of $2n$ arithmetic operations to determine $P_n(x)$ cannot be less.

Let us compare the algorithm based on Horner scheme with the following one:

$Compute$
$$x^2, x^3, \ldots, x^n,$$
and
$$P_n(x) = a_n x^n + a_{n-1} x^{n-1} + \ldots + a_1 x + a_0.$$

Clearly, to compute $x^2, x^3, \ldots, x^n$, we have to execute $n-1$ multiplications, and then, to compute $P_n(x)$, we need additionally n multiplications, and n additions. Thus, the total number of arithmetic operations involved in the second algorithm is 3n-1.

The above two ways of evaluation of $P_n(x)$ suggest that any numerical problem can be solved by a number of algorithms with different number of arithmetic operations involved in each, that is, with different computational complexity. Obviously, we apply optimal algorithm with minimum number of arithmetic operations, when such algorithm is known and produces satisfactory solution. However, in complex problems, optimal algorithm may be unknown or even may not exist, or the optimal algorithm is known but produces unsatisfactory solution because of its instability. In the case when optimal and stable algorithm is not available, we may use any stable algorithm with relatively small computational complexity.

There are two important properties which are expected from any good algorithm

1. *algorithm with the smallest computational complexity, that is, optimal algorithm with minimum number of arithmetic operations, or close to an optimal algorithm,*

2. *Stable algorithm, that is, algorithm which is resistance against any perturbation of input data or partial results*

caused be round-off errors.

As we have stated earlier, an algorithm is stable if in its implementation round-of errors accumulation is proportional to a computer precision.

We shall show that Horner scheme is an optimal and a stable algorithm. In order to prove that Horner scheme is the optimal algorithm, we apply the principle of mathematical induction. Namely, let the thesis $T(n)$ be

to evaluate

$$P_n(x) = a_n x^n + a_{n-1} x^{n-1} + \cdots + a_1 x + a_0,$$

we have to execute 2n multiplications and additions.

The thesis $T(1)$ is true, since

to find

$$P_1(x) = a_1 x + a_0,$$

we must execute one multiplication and one addition.

Now, assume that the thesis $T(n)$ is true *i.e.*

to compute

$$P_n(x) = a_n x^n + a_{n-1} x^{n-1} + \cdots + a_1 x + a_0$$

$$= (...((a_n x + a_{n-1}) x + a_{n-2}) x + \ldots + a_1) x + a_0,$$

we must execute 2n multiplications and additions.

If it is so, then $T(n)$ implies $T(n+1)$. Indeed, the polynomial

$$P_{n+1}(x) = a_{n+1} x^{n+1} + a_n x^n + \cdots + a_1 x + a_0$$

$$= (...(\underbrace{(a_{n+1} x + a_n) x + a_{n-1}) x + a_{n-2}) x + \ldots + a_1) x + a_0}_{P_n(x)}$$

By assumption, we may find $P_n(x)$ using $2n$ multiplications and $2n$ additions. Then, to evaluate $P_{n+1}(x)$, we additionally need one multiplication and one addition. Thus, the minimum number of operations required in Horner scheme is $2n + 2 = 2(n + 1)$.

Now, we shall investigate the round-off error in Horner scheme. Let x and a_i, $i = 0, 1, \ldots, n$ be computer floating point numbers and let ϵ_i and η_i be round-off errors associated with addition and multiplication, respectively.

Then, we have

$$
\begin{aligned}
fl(P_n(x)) = \ & (\cdots (a_n x(1 + \eta_n) + a_{n-1})(1 + \epsilon_{n-1})x(1 + \eta_{n-1}) \\
& + a_{n-2})(1 + \epsilon_{n-2})x(1 + \eta_{n-2} \\
& + a_{n-3})(1 + \epsilon_{n-3})x(1 + \eta_{n-3} \\
& + \cdots\cdots\cdots\cdots + \\
& + a_1(1 + \epsilon_1)x(1 + \eta_1) \\
& + a_0(1 + \epsilon_0).
\end{aligned}
$$

Hence, for $\eta_0 = 0$ and $\epsilon_n = 0$, we have

$$
\begin{aligned}
fl(P_n(x)) = \ & \\
& a_n x^n (1 + \eta_n)(1 + \eta_{n-1})\cdots(1 + \eta_1)(1 + \epsilon_{n-1})\cdots(1 + \epsilon_0) \\
& + a_{n-1} x^{n-1}(1 + \eta_{n-1})\cdots(1 + \eta_1)(1 + \epsilon_{n-1})\cdots(1 + \epsilon_0) \\
& + \cdots\cdots\cdots\cdots\cdots + \\
& + a_1 x(1 + \eta_1)(1 + \epsilon_1)(1 + \epsilon_0) \\
& + a_0(1 + \epsilon_0).
\end{aligned}
$$

and

$$
fl(P_n(x)) = \sum_{i=0}^{n} a_i x^i \prod_{j=0}^{i} (1 + \eta_j)(1 + \epsilon_j).
$$

Because $\mid \eta_i \mid \leq \delta$, $\eta_0 = 0$ and $\mid \epsilon_i \mid \leq \delta$, $i = 0, 1, \cdots, n$, therefore

$$|fl(P_n(x)) - P_n(x)| = |\sum_{i=0}^{n} a_i x^i [\prod_{j=0}^{i} (1 + \eta_j)(1 + \epsilon_j) - 1]|$$

$$\leq [(1 + \delta)^{2n} - 1] \sum_{i=0}^{n} |a_i x^i|$$

Let us note that $(1 + \delta)^{2n} - 1 \approx 2n\delta$ for a small δ, so that the absolute error is proportional to the computer precision, that is

$$|fl(P_n(x)) - P_n(x)| \approx 2n\delta \sum_{i=0}^{n} |a_i x^i|.$$

The relative error estimate is:

$$\frac{|fl(P_n(x)) - P_n(x))|}{|\sum_{i=0}^{n} a_i x^i|} \leq ((1 + \delta)^{2n} - 1) \frac{\sum_{i=0}^{n} |a_i x^i|}{|\sum_{i=0}^{n} a_i x^i|}$$

$$\leq 2n\delta \frac{\sum_{i=0}^{n} |a_i x^i|}{|\sum_{i=0}^{n} a_i x^i|}.$$

Example 1.5 *Consider the following polynomial:*

$$P_4(x) = 126.7x^4 + 34.82x^3 + 8.458x^2 + 0.2347x - 9554$$

(i). *Evaluate $P_4(2.875)$ in 4-digit floating point arithmetic using Horner scheme. Determine absolute and relative round-off errors.*

(ii). *Give an estimate of the absolute and relative round-off errors for evaluation of $P_n(x)$ by Horner scheme assuming that x is an 4-digit floating point number.*

Solution (i). The Horner's scheme of the polynomial $P_4(x)$ is

$$P_4(x) = (((126.7*x+34.82)*x+8.458)*x+0.2347)*x-9554$$

Rounding off after each operation in 4-digit floating point arithmetic, we obtain

$$
\begin{aligned}
fl(P_4(2.875)) \; &= fl((((126.7 * 2.875 + 34.82) * 2.875 + 8.458) \\
&\quad *2.875 + 0.2347) * 2.875 - 9554) \\
&= fl(((399.1 * 2.875 + 8.458) * 2.875 + 0.2347) \\
&\quad *2.875 - 9554) \\
&= fl((1155 * 2.875 + 0.2347) * 2.875 - 9554) \\
&= fl(3321 * 2.875 - 9554) = fl(9548 - 9554) \\
&= -6.
\end{aligned}
$$

and the exact value

$$
P_4(2.875) = 0.25.
$$

So that, the absolute round-off error

$$
\mid fl(P_4(2.875) - P_4(2.875) \mid = \mid -6 - 0.25 \mid = 6.25
$$

and the relative round-off error

$$
\mid \frac{fl(P_4(2.875)) - P_4(2.875)}{P_4(2.875)} \mid = \frac{6.25}{0.25} \approx 25.
$$

To evaluate a polynomial $P_n(x) = a_n x^n + a_{n-1} x^{n-1} + ... + a_0$ at a point x by Horner's scheme in **Mathematica**, we input data x and the list of coefficients $\{a_n, a_{n-1}, ..., a_0\}$. Then, we execute the following commands:

```
horner[x_,a_]:=Fold[(#1*x+#2)&, First[a],Rest[a]];
horner[x,a]
```

For example, to evaluate $P_4(x)$, we input data

```
x=2.875; a={126.7,34.82,8.458,0.2347,-9554};
```

Then, we define the function

```
horner[x_,a_]:=Fold[(#1*x+#2)&, First[a],Rest[a]];
```

Calling the function

```
horner[x_,a_]
```

we obtain $P_4(2.875) = 0.474172$

Following the details of the analysis of round-off error in the Horner's scheme, we may obtain the upper bounds of the round-off errors. Namely, we have

$$
\begin{aligned}
fl(P_4(x)) \quad &= ((((126.7x(1+\eta_4)+34.82)(1+\epsilon_3)x(1+\eta_3) \\
&+8.458)(1+\epsilon_2)x(1+\eta_2)+0.2347)(1+\epsilon_1)x(1+\eta_1) \\
&-9554(1+\epsilon_0).
\end{aligned}
$$

Hence, for $\eta_0 = 0$

$$
\begin{aligned}
fl(P_4(x)) \quad &= 126.7x^4(1+\eta_4)(1+\eta_3)(1+\eta_2)(1+\eta_1) \\
&*(1+\epsilon_3)(1+\epsilon_2)(1+\epsilon_1)(1+\epsilon_0) \\
&+34.82x^3(1+\eta_3)(1+\eta_2)(1+\eta_1)(1+\epsilon_3)(1+\epsilon_2) \\
&*(1+\epsilon_1)(1+\epsilon_0) \\
&+8.458x^2(1+\eta_2)(1+\eta_1)(1+\epsilon_2)(1+\epsilon_1)(1+\epsilon_0) \\
&+0.2347x(1+\eta_1)(1+\epsilon_1)(1+\epsilon_0) \\
&-9554(1+\epsilon_0) \\
&= \sum_{i=0}^{4} a_i x^i \prod_{j=0}^{i}(1+\eta_j)(1+\epsilon_j),
\end{aligned}
$$

where $a_0 = -9554$, $a_1 = 0.2347$, $a_2 = 8.458$, $a_3 = 34.82$, $a_4 = 126.7$.

Since $\mid \eta_i, \mid$, $\eta_0 = 0$ and

$$
\mid \epsilon_i \mid < \delta = \frac{1}{2}10^{-3} = 0.0005, \ i = 0, 1, 3, 4,
$$

we get the following estimation of the absolute round-off error

$$|fl(P_4(x)) - P_4(x)| \leq ((1+\delta)^8 - 1) \sum_{i=0}^{4} |a_i x^i|$$

$$\leq 8\delta \sum_{i=0}^{4} |a_i x^i| = 0.004 \sum_{i=0}^{4} |a_i x^i|.$$

and of the relative round-off error

$$\left| \frac{fl(P_4(x)) - P_4(x)}{P_4(x)} \right| \leq 0.004 \frac{\sum_{i=0}^{4} |a_i x^i|}{|\sum_{i=0}^{4} a_i x^i|}$$

for $P_4(x) \neq 0$.

1.6 Exercises

Question 1.1 *Calculate $fl(10^4 + 1.414213562)$ in 5-digit floating point arithmetic. Determine the absolute and relative round-off errors.*

Question 1.2 *Calculate in the 4-digit floating point arithmetic the following expressions:*

    (a) $43.65 + 126.8 + 0.9754,$

    (b) $54.09 * 0.8756 * 2.645,$

    (c) $\dfrac{4.876}{0.6543} - 23.65,$

    (d) $45.56 * 4.875 + 102.6 * 0.795 - 1251 * 346.4.$

Determine the absolute and relative round-off errors.

Question 1.3 *Let the numbers x and y be rounded-off in r-digit floating point arithmetic, i.e. $\bar{x} = fl(x)$, $\bar{y} = fl(y)$, $\Delta x = |x - fl(x)| \leq \epsilon$ and $\Delta y = |y - fl(y)| \leq \epsilon$, where ϵ is the round-off error. Suppose that in propagation of relative round-off errors $\bar{x} \approx x$ and $\bar{y} \approx y$.*
Show the following estimates of the relative round-off errors:

1. (a)

$$\frac{|\Delta(x+y)|}{|\overline{x}+\overline{y}|} \leq \max\{\frac{|\Delta x|}{|\overline{x}|}, \frac{|\Delta y|}{|\overline{y}|}\} \leq \max\{\frac{\epsilon}{|\overline{x}|}, \frac{\epsilon}{|\overline{y}|}\}.$$

provided that sign x = sign y.

(b)

$$\frac{|\Delta(xy)|}{|\overline{xy}|} \approx \frac{|\Delta x|}{|\overline{x}|} + \frac{|\Delta y|}{|\overline{y}|} \leq \frac{\epsilon}{|\overline{x}|} + \frac{\epsilon}{|\overline{y}|} \leq 2\delta.$$

Question 1.4 *Consider the following function:*

$$g(x, y, z) = xy + z \text{ for } x, y, z \in (-\infty, \infty).$$

(a). *Evaluate*

$$g(8.75, 4.86, 9.65)$$

in 3-digit floating point arithmetic.
(b). *Give an estimate of the absolute and relative round-off errors for evaluation of $g(x, y, z)$ assuming that $x, y, z \in (-\infty, \infty)$ are computer numbers and the computer precision $\delta = 0.005$.*

Question 1.5 *Give an estimate of the absolute and relative round-off errors of the following products*

$$(i) \quad \prod_{i=1}^{n} \frac{1}{x_i}, \quad (ii) \quad \sum_{i=1}^{n} a_i b_i,$$

where x_i , a_i and b_i , $i = 1, 2, \ldots, n$ are computer numbers with r-decimal digit mantissa.

Question 1.6 *Assume that the function*

$$G(x, y, z) = x^2 * y^2 * z^2, \quad x, y, z \in (-\infty, \infty),$$

has been evaluated in the floating point arithmetic with computer precision $\delta = \frac{1}{2}10^{1-r}$.

1. (a) *Give the estimates of the absolute and relative round-off errors of $fl(G(x, y, z))$.*

(b) Evaluate

$$G(1.25, 2.12, 4.50)$$

in the 3-digit floating point arithmetic. Determine the computer precision of the three digit arithmetic and estimate the absolute and relative round-off errors of the value $fl(G(1.25, 2.12, 4.50))$.

Question 1.7 *Consider the following polynomial:*

$$P_4(x) = 2x^4 + 2x^3 + 3x^2 + 4x + 5 \quad \text{for} \quad 0 \le x \le 1.$$

(a). *Evaluate $P_4(0.475)$ in 4-digit floating point arithmetic using Horner scheme. Determine absolute and relative round-off errors.*

(b). *Assume that $x \in [0, 1]$ is a 4-digit floating point number. Give an estimate of the absolute and relative round-off errors to evaluate $P_4(x)$ by Horner scheme in 4-digit floating point arithmetic.*

Question 1.8 *Give an estimate of the absolute and relative round-off errors for evaluation of the function*

$$s(x) = \sum_{i=1}^{n} \frac{i x^{i-1}}{3^i}$$

by Horner scheme in 3-digit floating point arithmetic. What is the error bound for $n = 5$, $x = 0.75$.

Question 1.9 *Consider the function*

$$f(x) = \frac{1 + \sin^2 x}{1 + \cos^2 x} \quad -\infty < x < \infty.$$

1. *(a) Compute the conditional number of the function $f(x)$ at a point $x \in (-\infty, \infty)$. and show that the conditional number does not exceed $\frac{3}{2}|x|$.*

   (b) Find all functions for which the conditional number is constant.

EXERCISES IN MATHEMATICA

In order to convert a rational number x to a real
number x, we execute the command:
 N[x] or N[x, r], where r is the number of digits.
For example
    x = 1/7; r = 20; N[x, r]
    0.14285714285714285714

In order to round off a real number x to the integer
closest to x, we execute the command:
                Round[x].
For example

                Round[12.5123]
                13

In order to round a number x to r decimal digits
after dot, we execute the command:
                N[Round[10^r x]/10^r];
For example

                x = 23.453879; r = 1;
                N[Round[10^r x]/10^r]
                23.5

The command Floor[x] gives the greatest integer
less than or equal to x.
For example
                Floor[-21.546]
                -22

The command Ceiling[x] gives the smallest integer
greater than or equal to x.
For example

```
Ceiling[-21.546]
-21
```

Question 1.
Let x = 432.7568.
(a) Round off x up to two digits after dot.
(b) Find the greatest integer less or equal to x.
(c) Find the smallest integer greater or equal to x.

Send Orders for Reprints to reprints@benthamscience.net

Lecture Notes in Numerical Analysis with Mathematica, 2014, 27-62   **27**

CHAPTER 2

Natural and Generalized Interpolating Polynomials

Abstract

In this chapter, Lagranges and Hermits interpolation by polynomials, by trigonometric polynomials, by Chebyshevs polynomials and by generalized polynomials spanned on Chebyshevs systems of coordinates are presented. Lagranges and Newtons formulas to find the interpolating polynomials are derived and clarified. Mathematica modules are designed to determine interpolating polynomials. Fundamental theorems on interpolation with the errors bounds are stated and proved. The application of the theorems has been elucidated by examples. The Chapter ends with a set of questions.

Keywords: Lagrange polynomials, Newton polynomials, Chebyshev polynomials.

2.1 Lagrange Interpolation

In this section, we shall describe Lagrange interpolation of a function using natural and generalized polynomials. In order to state the problem of Lagrange interpolation, we consider a partition of the bounded interval $[a, b]$ by the following points:

$$a = x_0 < x_1 < x_2 < \ldots < x_n = b.$$

Let us assume that to each point x_i a number y_i, $i = 0, 1, ..., n$, is assigned, where

$$y_0, \ y_1, \ \ldots, \ y_n$$

are measured or calculated values of a function

$$y = f(x).$$

In fact, $f(x)$ can only be given at the interpolating points x_i, $i = 0, 1, ..., n$, so that

$$y_i = f(x_i), \quad i = 0, 1, \ldots, n.$$

Now, let us state the problem of Lagrange interpolation (*cf.* [2,3,4,5,8, 10,12,13,14,15,18,23])

Find a polynomial

$$p_n(x) = a_0 + a_1 x + \cdots + a_n x^n$$

of degree at most n such that

$$p_n(x_i) = y_i, \quad i = 0, 1, \ldots, n. \tag{2.1}$$

The following theorem holds:

Theorem 2.1 *There exists a unique interpolating polynomial $p_n(x)$ to $f(x)$ which satisfies the Lagrange conditions (2.1)*

Proof. From the conditions of interpolation (2.1), we obtain the following linear system of algebraic equations:

$$a_0 + a_1 x_0 + a_2 x_0^2 + \cdots + a_n x_0^n = y_0$$

$$a_0 + a_1 x_1 + a_2 x_1^2 + \cdots + a_n x_1^n = y_1$$

$$\cdots \cdots \cdots \cdots \cdots \cdots \cdots \cdots \cdots \cdots \cdots$$

$$a_0 + a_1 x_n + a_2 x_n^2 + \cdots + a_n x_n^n = y_n$$

(2.2)

where $a_0, a_1, \cdots, a_n$ are unknown coefficients of the polynomial $p_n(x)$.

The matrix of this system of equations is non-singular since its determinant [1]

$$\begin{vmatrix} 1 & x_0 & x_0^2 & \cdots & x_0^n \\ 1 & x_1 & x_1^2 & \cdots & x_1^n \\ \cdots & \cdots & \cdots & \cdots & \cdots \\ 1 & x_n & x_n^2 & \cdots & x_n^n \end{vmatrix} = \prod_{i>j}^n (x_i - x_j)$$

is different from zero if $x_i \neq x_j$ for $i \neq j$. Therefore, the system of algebraic equations (2.2) has one solution. This solution determines the unique interpolating polynomial $p_n(x)$.

We may find polynomial $p_n(x)$ solving the system of linear equations (2.2). However, this system of linear equations can be ill-conditioned, so that, the round-off errors may strongly affect the final results. Therefore, we should look for more economic and stable methods to find the interpolating polynomial $p_n(x)$. We shall deal with such methods later. Now, let us state the problem of Lagrange interpolation for generalized polynomials.

Let

$$\phi_0(x), \phi_1(x), \ldots, \phi_n(x)$$

be a sequence of linearly independent functions given on the interval $[a, b]$.

---

[1] Vandermonde's determinant

A linear combination

$$P_n(x) = a_0\phi_0(x) + a_1\phi_1(x) + \cdots + a_n\phi_n(x)$$

is called generalized polynomial.

We shall assume that any generalized polynomial $P_n(x)$ cannot have more than n different roots in the interval $[a, b]$. Then, the sequence of functions

$$\phi_0(x), \phi_1(x), \ldots, \phi_n(x),$$

which satisfies this assumption is called *Chebyshev's system.*

Similarly, we state the problem of Lagrange interpolation for generalized polynomials

Find a generalized polynomial

$$P_n(x) = a_0\phi_0(x) + a_1\phi_1(x) + \cdots + a_n\phi_n(x)$$

which satisfies the following conditions:

$$P_n(x_i) = y_i, \quad i = 0, 1, \ldots, n.$$

Below, we present examples of Chebyshev's systems.

2.2   Certain Chebyshev's Systems

Let us list the most common used basis of interpolating polynomials.

(a). Base of natural polynomials. Let

$$\phi_i(x) = x^i, \quad i = 0, 1, \ldots, n.$$

Then, the natural form of a polynomial is:

$$P_n(x) = a_0 + a_1 x + a_2 x^2 + \ldots + a_n x^n.$$

where $-\infty < x < \infty, \quad i = 0, 1, \cdots, n.$

(b). Lagrange base of polynomials. Let us consider

polynomials $l_i(x)$, $i = 0, 1 \ldots, n$, of degree n which satisfy the following conditions of interpolation:

$$l_i(x_j) = \begin{cases} 1 & \text{if } i = j, \\ 0 & \text{if } i \neq j. \end{cases} \qquad (2.3)$$

Clearly, the above conditions satisfy the following Lagrange polynomials:

$$l_i(x) = \frac{(x - x_0)(x - x_1) \cdots (x - x_{i-1})(x - x_{i+1}) \cdots (x - x_n)}{(x_i - x_0)(x_i - x_1) \cdots (x_i - x_{i-1})(x_i - x_{i+1}) \cdots (x_i - x_n)},$$

where $-\infty < x < \infty$, $i = 0, 1, \ldots, n$.

Thus, the Lagrange interpolating polynomial takes the form

$$P_n(x) = a_0 l_0(x) + a_1 l_1(x) + \ldots + a_n l_n(x).$$

Indeed, by conditions (2.3), we have

$$P_n(x_i) = a_i = y_i, \quad i = 0, 1, \cdots, n,$$

and

$$P_n(x) = y_0 l_0(x) + y_1 l_1(x) + \cdots + y_n l_n(x). \qquad (2.4)$$

Example 2.1 *Find the Lagrange interpolating polynomial which attains values 4,2,-1,10 at points -2,1,2,4. Write down the natural form of this polynomial.*

Solution. From the Lagrange formula (2.4), we have

$$\begin{aligned} P_3(x) &= 4\frac{(x - 1)(x - 2)(x - 4)}{(-2 - 1)(-2 - 2)(-2 - 4)} \\ &+ 2\frac{(x + 2)(x - 2)(x - 4)}{(1 + 2)(1 - 2)(1 - 4)} - 1\frac{(x + 2)(x - 1)(x - 4)}{(2 + 2)(2 - 1)(2 - 4)} \\ &+ 10\frac{(x + 2)(x - 1)(x - 2)}{(4 + 2)(4 - 1)(4 - 2)} \\ &= \frac{41}{72}x^3 - \frac{83}{72}x^2 - \frac{127}{36}x + \frac{55}{9}. \end{aligned}$$

In `Mathematica`, we find the polynomial $P_3(x)$ by the following command:

```
Expand[InterpolatingPolynomial[{{-2,4},{1,2},
                        {2,-1},{4,10}},x]]
```

(c). Newton's base of interpolating polynomials. Let us consider the following polynomials:

$$\phi_0(x) = 1$$
$$\phi_1(x) = x - x_0$$
$$\phi_2(x) = (x - x_0)(x - x_1)$$
$$\phi_3(x) = (x - x_0)(x - x_1)(x - x_2)$$
$$\cdots\cdots\cdots\cdots\cdots$$
$$\phi_n(x) = (x - x_0)(x - x_1)(x - x_2)\cdots(x - x_{n-1})$$

Then, Newton interpolating polynomial takes the following form:

$$P_n(x) = a_0 + a_1(x - x_0) + a_2(x - x_0)(x - x_1) + \cdots$$
$$+ \ a_n(x - x_0)(x - x_1)\cdots(x - x_{n-1}),$$

where the coefficients $a_0, a_1, \ldots, a_n$, are determined by the following conditions of interpolation (2.1):

$$a_0 = y_0$$

$$a_0 + a_1(x_1 - x_0) = y_1$$

$$a_0 + a_1(x_2 - x_0) + a_2(x_2 - x_0)(x_2 - x_1) = y_2$$

$$\cdots\cdots\cdots\cdots\cdots\cdots\cdots\cdots\cdots\cdots\cdots\cdots\cdots\cdots\cdots \quad \cdots$$

$$a_0 + a_1(x_n - x_0) + \cdots + a_n(x_n - x_0)\cdots(x_n - x_{n-1}) = y_n$$

The solution of the this system of linear equations is expressed in terms of the following finite differences:

$$a_0 = [x_0] = y_0$$

$$a_1 = [x_0 x_1] = \frac{[x_1] - [x_0]}{x_1 - x_0}$$

$$a_2 = [x_0 x_1 x_2] = \frac{[x_1 x_2] - [x_0 x_1]}{x_2 - x_0}$$

$$\cdots \cdots \cdots \cdots \cdots \cdots$$

$$a_n = [x_0 x_1 x_2 \cdots x_n] = \frac{[x_1 x_2 \cdots x_n] - [x_0 x_1 \cdots x_{n-1}]}{x_n - x_0}$$

$$(2.5)$$

Thus, Newton's interpolating polynomial is:

$$P_n(x) = [x_0] + [x_0 x_1](x - x_0) + [x_0 x_1 x_2](x - x_0)(x - x_1) + \cdots$$

$$+ [x_0 x_1 x_2 \cdots x_n](x - x_0)(x - x_1) \cdots (x - x_{n-1}).$$

$$(2.6)$$

In order to compute the coefficients $a_0, a_1, \cdots, a_n$ from the formulas (2.5), it is convenient to use the following table:

x_0	y_0				
		$[x_0 x_1]$			
x_1	y_1		$[x_0 x_1 x_2]$		
		$[x_1 x_2]$		$[x_0 x_1 x_2 x_3]$	
x_2	y_2		$[x_1 x_2 x_3]$		$[x_0 x_1 x_2 x_3 x_4]$
		$[x_2 x_3]$		$[x_1 x_2 x_3 x_4]$	
x_3	y_3		$[x_2 x_3 x_4]$		
		$[x_3 x_4]$			
$\vdots$	$\vdots$				
x_{n-1}	y_{n-1}				
		$[x_{n-1} x_n]$			
x_n	y_n				

Example 2.2 *Find the Newton's interpolating polynomial*

which attains values 1,3,2 and 5 at the points 0,2,3,5. Reduce this polynomial to its natural form.

By the Newton's formula (2.6), we have

$$P_3(x) = 1 + [02](x - 0) + [023](x - 0)(x - 2)$$
$$+ \ [0235](x - 0)(x - 2)(x - 3).$$

We can get the coefficients **[02]**, **[023]** and **[0235]** from the following table:

$$
\begin{array}{c|c}
0 & 1 \\
 & \quad 1 \\
2 & 3 \qquad -\dfrac{2}{3} \\
 & \quad -1 \qquad \dfrac{3}{10} \\
3 & 2 \qquad \dfrac{5}{6} \\
 & \quad \dfrac{3}{2} \\
5 & 5
\end{array}
$$

Hence, the Newton's interpolating polynomial is

$$P_3(x) = 1 + x + (-\frac{2}{3}x(x - 2) + \frac{3}{10}x(x - 2)(x - 3)$$
$$= 1 + \frac{62}{15}x - \frac{13}{6}x^2 + \frac{3}{10}x^3.$$

To find the Newton's interpolating polynomial $P_3(x)$ in
Mathematica, we execute the following command:

```
InterpolatingPolynomial[{{0,1},{2,3},{3,2},
                {5,5}},x]
```

We obtain the natural form of the polynomial $P_3(x)$ by the
command

```
Expand[InterpolatingPolynomial[{{0,1},{2,3},{3,2},
                {5,5}},x]]
```

(d). Base of piecewise linear splines. Let us consider the following piecewise linear splines

$$\psi_i(x) = \begin{cases} 0 & \text{if } x \le x_{i-1}\, or\, x \ge x_{i+1}, \\[2mm] \dfrac{x - x_{i-1}}{x_i - x_{i-1}} & \text{if } x_{i-1} \le x \le x_i, \\[2mm] \dfrac{x_{i+1} - x}{x_{i+1} - x_i} & \text{if } x_i \le x \le x_{i+1}, \end{cases} \qquad (2.7)$$

for $i = 1, 2, ..., n - 1$,

$$\psi_0(x) = \begin{cases} 0 & \text{if } x > x_1, \\[2mm] \dfrac{x_1 - x}{x_1 - x_0} & \text{if } x_0 \le x_1. \end{cases}$$

and

$$\psi_n(x) = \begin{cases} 0 & \text{if } x < x_{n-1}, \\[2mm] \dfrac{x - x_{n-1}}{x_n - x_{n-1}} & \text{if } x_{n-1} \le x \le x_n. \end{cases}$$

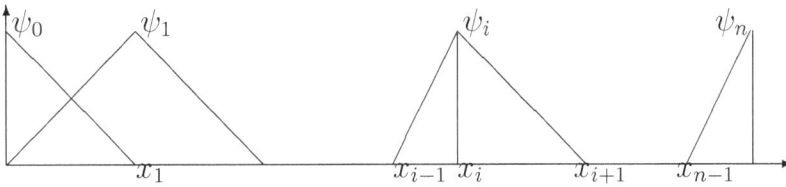

Fig. 2.1. Linear spline $\psi_i(x)$

Since

$$\psi_i(x) = \begin{cases} 1 & \text{if } i = j, \\ 0 & \text{if } i \ne j, \end{cases}$$

we get

$$a_i = y_i, \quad i = 0, 1, \ldots, n.$$

Hence, the linear piecewise interpolating polynomial (see **Fig. 2.1**) takes the following form:

$$P_1(x) = y_0 \psi_0(x) + y_1 \psi_1(x) + \cdots + y_n(x) \psi_n(x). \qquad (2.8)$$

Example 2.3 *Find a piecewise linear interpolating spline which attains values 1,3,0, and 4 at points 0,1,2 and 3.*

Following the formula (2.8), we find

$$P_1(x) = 1\psi_0(x) + 3\psi_1(x) + 0\psi_2(x) + 4\psi_3(x) =$$

$$= \begin{cases} \dfrac{1-x}{1-0} + 3\dfrac{x-0}{1-0} & \text{if } 0 \le x \le 1, \\[2ex] 3\dfrac{2-x}{2-1} + & \text{if } 1 \le x \le 2, \\[2ex] 4\dfrac{x-2}{3-2} & \text{if } 2 \le x \le 3, \\[2ex] 0 & \text{otherwise,} \end{cases}$$

$$= \begin{cases} 1+2x & \text{if } 0 \le x \le 1, \\ 6-3x & \text{if } 1 \le x \le 2, \\ 4x-8 & \text{if } 2 \le x \le 3, \\ 0 & \text{otherwise.} \end{cases}$$

We present the piecewise linear spline on the following figure (see **Fig. 2.2.**)

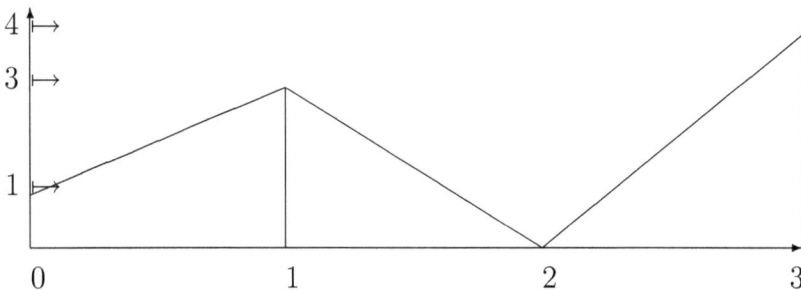

Fig. 2.2. $P_1(x)$.

(e) Chebyshev's polynomials. Let us consider Chebyshev's polynomials

$T_0(x) = 1,$

$T_1(x) = x,$

$T_2(x) = 2x^2 - 1,$

$\cdots\cdots\cdots\cdots\cdots,$

$T_n(x) = 2x\, T_{n-1}(x) - T_{n-2}(x), \quad -\infty < x < \infty, \quad n = 2, 3, \ldots.$
or

$T_n(x) = cos(n\, arc\, cos\, x), \qquad -\infty < x < \infty, \quad n = 0, 1, \ldots;$

The interpolating polynomial spaned by Chebyshev's polynomials is

$P_n(x) = a_0 T_0(x) + a_1 T_1(x) + \cdots + a_n T_n(x), \quad -\infty < x < \infty.$

Example 2.4 *Find an interpolating polynomial spaned by Chebyshev's polynomials which attains values 1,0,1 at points 0,1 and 2. Reduce this polynomial to its natural form.*

Thus, we have

$$P_2(x) = a_0 T_0(x) + a_1 T_1(x) + a_2 T_2(x),$$

where the coefficients a_0, a_1, a_2 are determined by the equations

$$a_0 T_0(0) + a_1 T_1(0) + a_2 T_2(0) = 1,$$
$$a_0 T_0(1) + a_1 T_1(1) + a_2 T_2(1) = 0,$$
$$a_0 T_0(2) + a_1 T_1(2) + a_2 T_2(2) = 1,$$

which take the following form:

$$a_0 - a_2 \qquad\quad = 1$$

$$a_0 + a_1 + a_2 \quad = 0$$

$$a_0 + 2a_1 + 7a_2 = 1.$$

Hence, we find

$$a_0 = \frac{3}{2}, \quad a_1 = -2, \quad a_2 = \frac{1}{2},$$

and

$$P_2(x) = \frac{3}{2} - 2x + \frac{1}{2}(2x^2 - 1) = 1 - 2x + x^2.$$

To find the Chebyshev's interpolating polynomial $P_2(x)$ we execute the following program:

```
m=Table[ChebyshevT[n,x],{x,0,2},{n,0,2}];
b={1,0,1};
a={a1,a2,a3};
a=LinearSolve[m,b];
a[[1]]-a[[2]]x+a[[3]](2x^2-1)
```

2.3 Relationship Between a Function and its Interpolating Polynomial

In the previous section, we have shown that there is only one interpolating polynomial $p_n(x)$ spaned by the points

$$x_0 < x_1 < \cdots < x_n,$$

which satisfies the conditions: $y_i = f(x_i), \quad i = 0, 1, \ldots n$. As we have seen, this unique polynomial $p_n(x)$ can be given by different formulas, (Lagrange or Newton's formulas). Obviously, it is interesting to know what is a relationship between the function $y = f(x)$ and its unique interpolating polynomial $p_n(x)$. This relationship is given in the following theorem:

Theorem 2.2 *If $y = f(x)$ is a function $(n+1)$ times continuously differentiable in the interval $[a,b]$, then there exists a point $\xi_x \in (a, b)$ such that*

$$f(x) = p_n(x) + \frac{f^{(n+1)}(\xi_x)}{(n+1)!}(x - x_0)(x - x_1) \cdots (x - x_n)$$

for all $x \in [a, b]$.

Proof. Let us consider the following auxiliary function

$$g(t) = p_n(t) - f(t) - \frac{w_{n+1}(t)}{w_{n+1}(x)}[p_n(x) - f(x)], \qquad (2.9)$$

for $x \in [a, b]$, $x \neq x_i$, $i = 0, 1, \ldots, n$, where

$$w_{n+1}(x) = (x - x_0)(x - x_1) \cdots (x - x_n). \qquad (2.10)$$

Since $f(x)$ is $(n + 1)$ times continuously differentiable in the interval $[a, b]$, $g(t)$ is also $(n + 1)$ times continuously differentiable in the interval $[a, b]$. From the formula (2.9), it follows that g(t) possesses $(n + 2)$ different zeros in the interval $[a, b]$.
Then, we have

$$g(x) = 0 \quad \text{and} \quad g(x_i) = 0, \quad i = 0, 1, \ldots, n.$$

By the Rolle's theorem, the derivative $g'(t)$ has $(n + 1)$ different zeros in the open interval (a, b). Therefore, its second derivative $g''(t)$ has n different zeros in the interval (a, b). In this way, we can conclude that the derivative $g^{(n+1)}(t)$ of order $(n + 1)$ possesses at least one zero in the interval (a, b). Therefore, there exists a point $\xi_x \in (a, b)$ such that

$$g^{(n+1)}(\xi_x) = 0. \qquad (2.11)$$

On the other hand, by (2.11), we have

$$g^{(n+1)}(\xi_x) = -f^{(n+1)}(\xi_x) - \frac{(n + 1)!}{w_{n+1}(x)}[p_n(x) - f(x)] = 0.$$

Hence

$$f(x) = p_n(x) + \frac{f^{(n+1)}(\xi_x)}{(n + 1)!}(x - x_0)(x - x_1) \cdots (x - x_n).$$

Error of interpolation. From the theorem, it follows that the error of interpolation

$$f(x) - p_n(x)$$

satisfies the equality

$$f(x)-p_n(x) = \frac{f^{(n+1)}(\xi_x)}{(n+1)!}(x-x_0)(x-x_1)\cdots(x-x_n). \quad (2.12)$$

for certain $\xi_x \in (a,b)$.

Hence

$$\mid f(x) - p_n(x) \mid \leq \frac{M^{(n+1)}}{(n+1)!}\mid(x-x_0)(x-x_1)\cdots(x-x_n)\mid,$$

where

$$M^{(n+1)} = \max_{a \leq x \leq b} \mid f^{(n+1)}(x) \mid .$$

Now, let us note that

$$\mid (x-x_0)(x-x_1)\cdots(x-x_n) \mid \leq \frac{n!h^{n+1}}{4},$$

for $n \geq 1$, and $h = \max_{0 \leq i \leq n}(x_{i+1} - x_i)$.

Indeed, we have

$$\mid (x-x_i)(x-x_{i+1}) \mid \leq \frac{h^2}{4}, \qquad x \in [x_i, x_{i+1}],$$

and

$$\mid x - x_{i+k} \mid \leq (k+1)h, \qquad k = 0, 1, \ldots, n-i.$$

Hence

$$\mid (x-x_0)(x-x_1)\cdots(x-x_i)(x-x_{i+1})\cdots(x-x_n) \mid \leq \frac{n!h^{n+1}}{4}.$$
$$(2.13)$$

Combining the above two inequalities, we obtain the following error estimate:

$$\mid f(x) - p_n(x) \mid \leq \frac{M^{(n+1)}}{4(n+1)}h^{n+1}, \qquad \text{when} \quad a \leq x \leq b.$$
$$(2.14)$$

Example 2.5 *Estimate the error of interpolation of a function twice continuously differentiable in the interval $[a,b]$ by piecewise linear spline.*

In this case, the interpolating generalized polynomial is

$$P_n(x) = f(x_0)\psi_0(x) + f(x_1)\psi_1(x) + \cdots + f(x_n)\psi_n(x).$$

Let $x_i < x < x_{i+1}$. Then, by formula (2.12), we have

$$f(x) - P_n(x) = \frac{f''(\xi_x)}{2}(x - x_i)(x - x_{i+1}), \qquad x_i < x < x_{i+1}.$$

Hence

$$| f(x) - P_n(x) | \leq \frac{M^{(2)}}{2} | (x - x_i)(x - x_{i+1}).$$

Since

$$| (x - x_i)(x - x_{i+1}) | \leq \frac{h^2}{4}, \qquad x_i < x < x_{i+1},$$

the error of interpolation satisfies the following inequality

$$| f(x) - P_n(x) | \leq \frac{M^{(2)}}{8} h^2, \qquad a \leq x \leq b.$$

Let us note that the constant $\frac{1}{8}$ is possible, that is, if we consider any estimate of the form

$$| f(x) - P_n(x) | \leq C M^{(2)} h^2,$$

then the constant C must be greater than or equal to $\frac{1}{8}$. Indeed, for the function $f(x) = \frac{1}{2}x(1 - x)$, $0 \leq x \leq 1$ and $h = 1$, we have $P_n(x) = 0$ and

$$| f(x) - P_n(x) | = \frac{1}{8} \quad \text{for} \quad x = \frac{1}{2}.$$

Example 2.6 *Function $f(x) = \cosh(\frac{x}{2})$ is tabulated below.*

x	-2	-1	1	2
y	1.54308	1.12763	1.12763	1.54308

(a) *Using data in the table, find Lagrange interpolating polynomial $P(x)$ and reduce it to the natural form.*

(b) *Construct a divided difference table based on the above data and determine Newton's interpolating polynomial.*

(c) *Interpolate the value $f(0.8)$ and show that the error of interpolation satisfies the following inequality*

$$| f(x) - P(x) | < 0.0161 \quad \text{for each} \ \ x \in [-2, 2].$$

Solution of (a). From the Lagrange formula

$$
\begin{aligned}
P_3(x) \ &= 1.54308 \frac{(x+1)(x-1)(x-2)}{(-2+1)(-2-1)(-2-2)} \\
&+1.12763 \frac{(x+2)(x-1)(x-2)}{(-1+2)(-1-1)(-1-2)} \\
&+1.12763 \frac{(x+2)(x+1)(x-2)}{(1+2)(1+1)(1-2)} \\
&+1.54308 \frac{(x+2)(x+1)(x-1)}{(2+2)(2+1)(2-1)}.
\end{aligned}
$$

Hence, the natural form of the interpolating polynomial is

$$P_2(x) = 0.1384834x^2 + 0.9891464.$$

Solution of (b). By the Newton's formula

$$
\begin{aligned}
P_2(x) \ &= 1.54308 + [-2, -1](x+2) + [-2, -1, 1](x+2)(x+1) \\
&+[-2, -1, 1, 2](x+2)(x+1)(x-1).
\end{aligned}
$$

We can get the coefficients $[-2,-1]$, $[-2,-1,1]$ and $[-2,-1,1,2]$ from the following table

$$
\begin{array}{r|llll}
-2 & 1.54308 & & & \\
 & & -0.41545 & & \\
-1 & 1.12763 & & 0.13848 & \\
 & & 0 & & 0 \\
1 & 1.12763 & & 0.13848 & \\
 & & 0.41545 & & \\
2 & 1.54308 & & &
\end{array}
$$

Hence, the Newton's interpolating polynomial is

$$P_2(x) = 1.54308 - 0.41545(x + 2) + 0.13848(x + 2)(x + 1).$$

In Mathematica, we can obtain the natural form of the interpolating polynomial $P_2(x)$ by the following program:

```
data={{-2,Cosh[-2/2]},{-1,Cosh[-1/2]},
{1,Cosh[1/2]},{2,Cosh[2/2]}};
N[Expand[InterpolatingPolynomial[data,x]],4]
```

Solution of (c). The interpolating value of $f(0.8) \approx P(0.8) = 1.077759$. In order to estimate the error of interpolation, we apply the formula

$$E(x) = f(x) - P_n(x) = \frac{f^{(n+1)}(\xi_x)}{(n + 1)!}(x - x_0)(x - x_1) \ldots (x - x_n).$$

Hence, for $n = 3$, $h = 2$, by (2.14), we have

$$E(x) = cosh(\tfrac{x}{2}) - (1.54308 - 0.41545(x + 2)$$

$$+0.13848(x + 2)(x + 1))$$

$$= \frac{cosh(\xi_x/2)}{16 * 4!}(x + 2)(x + 1)(x - 1)(x - 2)$$

$$\leq \frac{cosh(\xi_x/2)}{16 * 4 * (3 + 1)}2^4 \leq 0.0964.$$

However, using the inequality

$$|(x + 2)(x + 1)(x - 1)(x - 2)| \leq 4 \quad \text{for all} \ \ x \in [-2, 2],$$

we obtain more optimistic error estimate

$$|E(x)| \leq 0.0161$$

for all $x \in [-2, 2]$.

2.4   Optimal Interpolating Polynomial

As we know, the error of interpolation

$$E(x) = f(x) - P_n(x) = \frac{f^{(n+1)}(\xi_x)}{(n + 1)!}w_{n+1}(x),$$

where $w_{n+1}(x) = (x - x_0)(x - x_1)...(x - x_n)$.
Clearly, the error $E(x)$ depends on distribution of the interpolating points $x_0, x_1, \ldots, x_n$, in the interval $a, b]$. Thus, $E(x)$ attains its minimum if

$$\max_{a \leq x \leq b} |w_{n+1}(x)| = \min_{x_0, x_1, \ldots x_n} \max_{a \leq x \leq b} |w_{n+1}(x)|, \qquad (2.15)$$

From formula (2.15), it follows that the maximal error of interpolation attains its smallest value if $w_{n+1}(x)$ is the best polynomial among all polynomials with the coefficient

$a_{n+1} = 1$ at x^{n+1}, approximating the zero-function on the interval $[a, b]$. Then, points $x_0, x_1, \ldots, x_n$ which satisfy the formula (2.15) are called *Chebyshev's interpolating knots* and the interpolating polynomial $P_n(x)$ spaned by the Chebyshev's knots is called *optimal interpolating polynomial*.

In 1857, E.W. Chebyshev found that

$$w_{n+1}(x) = \tilde{T}_{n+1}(x) = \frac{1}{2^n} cos((n+1) \; arc \; cos \; x)$$

is the best polynomial approximating the function identically equal to zero on the interval $[-1, 1]$. The Chebyshev polynomials $\tilde{T}_{n+1}(x)$, $\quad n = 0, 1, \cdots$; satisfy the following recursive formula:

$$\tilde{T}_0(x) = 2,$$
$$\tilde{T}_1(x) = x,$$
$$\tilde{T}_2(x) = x^2 - \tfrac{1}{2},$$
$$\cdots\cdots\cdots\cdots$$
$$\tilde{T}_{n+1}(x) = x\tilde{T}_n(x) - \tfrac{1}{4}\tilde{T}_{n-1}(x).$$

Thus, the optimal interpolating polynomial for a function $f(x)$ on the interval $[-1, 1]$ is spaned by the roots of the Chebyshev's polynomial $T_{n+1}(x)$. Solving the equation

$$cos((n+1) \; arc \; cos \; x) = 0,$$

we find the roots

$$x_k = cos\frac{(2k+1)\pi}{2(n+1)}, \quad k = 0, 1, \ldots, n.$$

Since

$$|\tilde{T}_{n+1}(x)| = |(x - x_0)(x - x_1) \cdots (x - x_n)| \le \frac{1}{2^n},$$

we arrive at the following error estimate:

$$|f(x) - P_n(x)| \le \frac{M^{(n+1)}}{2^n(n+1)!}, \quad -1 \le x \le 1. \quad (2.16)$$

In order to determine an optimal interpolating polynomial on the interval $[a, b] \neq [-1, 1]$, we may use the following linear mapping:

$$z = \frac{1}{b - a}(2x - a - b),$$

to transform the interval $[a, b]$ on the interval $[-1.1]$.

Example 2.7 *Find an optimal interpolating polynomial of degree at most two for the following functions*

$$\textbf{(i)} \quad f(x) = \cos\tfrac{\pi}{2}x, \quad -1 \leq x \leq 1,$$

$$\textbf{(ii)} \quad f(x) = \sqrt{1 + x}, \quad 0 \leq x \leq 2.$$

Estimate the error of interpolation.

Solution (i). For $n = 2$, the Chebyshev's interpolating knots are

$$x_0 = -\frac{\sqrt{3}}{2}, \quad x_1 = 0, \quad x_2 = \frac{\sqrt{3}}{2}.$$

Therefore

$$y_0 = f(x_0) = \cos\left(-\frac{\pi\sqrt{3}}{4}\right) = 0.208897,$$

$$y_1 = f(x_1) = \cos 0 = 1,$$

$$y_2 = f(x_2) = \cos\frac{\pi\sqrt{3}}{2} = 0.208897.$$

Using the Lagrange formula, we find

$$P_2(x) = .2089\frac{x(x - \frac{\sqrt{3}}{2})}{\frac{\sqrt{3}}{2}(\frac{\sqrt{3}}{2} + \frac{\sqrt{3}}{2})} + \frac{(x + \frac{\sqrt{3}}{2})(x - \frac{\sqrt{3}}{2})}{(0 + \frac{\sqrt{3}}{2})(0 - \frac{\sqrt{3}}{2})}$$

$$+ .2089\frac{x(x + \frac{\sqrt{3}}{2})}{(\frac{\sqrt{3}}{2} + \frac{\sqrt{3}}{2})\frac{\sqrt{3}}{2}}.$$

Hence, we get the natural form of the optimal interpolating polynomial

$$P_2(x) = -1.0548\,x^2 + 1.$$

From (2.16), we can get the following error estimate:

$$|f(x) - P_2(x)| = \left|\cos\frac{\pi}{2}x + 1.0548x^2 - 1\right| \le \frac{M^{(3)}}{3!\,4}.$$

Since

$$f'''(x) = \frac{\pi^3}{8}\,\sin\frac{\pi}{2}x,$$

we get

$$M^{(3)} = \frac{\pi^3}{8}max\,\left|\sin\frac{\pi}{2}x\right| = \frac{\pi^3}{8}.$$

Thus, the error of interpolation satisfies the inequality

$$\left|\cos\frac{\pi}{2}x + 1.0548\,x^2 - 1\right| \le \frac{\pi^3}{8\,3!\,4} = 0.161.$$

In `Mathematica`, we can find the optimal interpolating polynomial $P_2(x)$ by the following program:

```
data={{-Sqrt[3]/2,Cos[-Pi*Sqrt[3]/4]},{0,1},
{Sqrt[3]/2,Cos[Pi*Sqrt[3]/4]}};
N[Expand[InterpolatingPolynomial[data,x]],4]
```

Solution (ii). At first, let us transform the interval $[0, 2]$ on the interval $[-1, 1]$ using the following mapping:

$$z = x - 1.$$

Now, we shall interpolate the function

$$g(z) = f(z+1) = \sqrt{2+z}, \qquad -1 \le z \le 1.$$

Also, in this case, the Chebyshev's interpolating knots are:

$$z_0 = -\frac{\sqrt{3}}{2}, \qquad z_1 = 0, \qquad z_2 = \frac{\sqrt{3}}{2}.$$

Therefore

$$y_0 = g(z_0) = \sqrt{2 - \tfrac{\sqrt{3}}{2}} = 1.06488,$$

$$y_1 = g(z_1) = \sqrt{2} = 1.4142414,$$

$$y_2 = g(z_2) = \sqrt{2 + \tfrac{\sqrt{3}}{2}} = 1.931852.$$

Using the Lagrange formula, we find

$$P_2(z) = \sqrt{2 - \frac{\sqrt{3}}{2}} \; \frac{z(z - \tfrac{\sqrt{3}}{2})}{\tfrac{\sqrt{3}}{2}(\tfrac{\sqrt{3}}{2} + \tfrac{\sqrt{3}}{2})} + \sqrt{2} \frac{(z + \tfrac{\sqrt{3}}{2})(z - \tfrac{\sqrt{3}}{2})}{(0 + \tfrac{\sqrt{3}}{2})(0 - \tfrac{\sqrt{3}}{2})}$$

$$+ \sqrt{2 + \frac{\sqrt{3}}{2}} \; \frac{z(z + \tfrac{\sqrt{3}}{2})}{(\tfrac{\sqrt{3}}{2} + \tfrac{\sqrt{3}}{2})\tfrac{\sqrt{3}}{2}}.$$

Hence, after simplification

$$P_2(z) = \frac{1}{3\sqrt{2}}[6 + (\sqrt{3(4 + \sqrt{3})}$$

$$- \sqrt{4 - 3(\sqrt{3})})z + 2(\sqrt{3(4 + \sqrt{3})} - \sqrt{3(4 - \sqrt{3})})z^2].$$

Coming back to the original variable x, we obtain the following optimal interpolating polynomial

$$P_2(x) = 1.00453 + 0.456757x - 0.0470738x^2.$$

The error of interpolation satisfies the following inequality:

$$|f(x) - P_2(x)| \le \frac{M^{(3)}}{3! \, 2^2}.$$

Since

$$|f'''(x)| = \frac{3}{8}(1 + x)^{-\frac{5}{2}},$$

we have

$$M^{(3)} = max|f'''(x)| = \frac{3}{8},$$

and the error estimate

$$|f(x) - P_2(x)| \le \frac{3}{8 \, 3! \, 4} = 0.0156.$$

In $\mathtt{Mathematica}$, we can find the optimal interpolating polynomial $P_2(x)$ by the following program:

```
Clear[x,z];
z0=-Sqrt[3]/2;
z1=0;
z2=Sqrt[3]/2;
{{z0,Sqrt[2+z0]},{z1,Sqrt[2+z1]},{z2,Sqrt[2+z2]}};
P2=InterpolatingPolynomial[data,z];
z=x-1;
P2=Expand[P2,4];
```

2.5   Hermite Interpolation.

Within this section, we shall consider the Hermite's interpolation of a given data by a polynomial. This kind of interpolation requires as input data $y_0, y_1, \ldots, y_n$ and the derivatives $y_i', y_i'', \ldots, y_i^{(s_i-1)}$; $i = 0, 1, \ldots, s_i - 1$; at the interpolating knots $x_0, x_1, \ldots, x_n$.

Let us state Hermite interpolation problem as follows:

Find a polynomial

$$P_m(x) = a_0 + a_1 x + \cdots + a_m x^m,$$

which satisfies the following interpolating conditions:

$$P_m(x_i) = a_0 + a_1 x_i + a_2 x_i^2 + \cdots + a_m x_i^m = y_i,$$

$$P_m'(x_i) = a_1 + 2a_2 x_i + 3a_3 x_i^2 + \cdots + m a_m x_i^{m-1} = y_i',$$

$$P_m''(x_i) = 2a_2 + 3*2a_3 x_i + \cdots + m(m-1)a_m x_i^{m-2} = y_i'',$$

$$\ldots\ldots\ldots\ldots\ldots\ldots\ldots\ldots\ldots\ldots\ldots\ldots\ldots\ldots\ldots\ldots\ldots\ldots$$

$$P_m^{(s_i-1)}(x_i) = 2*3*\cdots*(s_i-1)x a_{s_i} + \cdots$$

$$+m(m-1)(m-2)\cdots(m-s_i+1)x_i^{m-s_i+2} a_m = y_i^{(s_i-1)},$$

for $i = 0, 1, \ldots, n;$ and $s_0 + s_1 + \cdots + s_n = m + 1$.
Under these conditions, there exists a unique Hermite interpolating polynomial $P_m(x)$. This can be shown in a similar way as it has been proved for the Lagrange interpolating polynomial. Let us note that the Hermite interpolation problem reduces to the Lagrange interpolation problem, if $s_0 = s_1 = s_2 \ldots = s_n = 1,$.
Below, we shall determine a Hermite interpolating polynomial in the case when $s_0 = s_1 = \cdots = s_n = 2$ and $m = 2n + 1$.
We shall first find a polynomial

$$P_{2n+1}(x) = a_0 + a_1 x + \cdots + a_{2n+1} x^{2n+1},$$

which satisfies the interpolating conditions

$$P_{2n+1}(x_i) = y_i, \quad P'_{2n+1}(x_i) = y'(x_i), \quad i = 0, 1, \ldots, n.$$

This polynomial, we shall find in the following form:

$$H_{2n+1}(x) = L_n(x) + w_{n+1}(x) H_n(x), \qquad (2.17)$$

where $L_n(x)$ is the Lagrange interpolating polynomial (2.4) and $w_{n+1}(x)$ is the polynomial given by formula (2.10). The polynomial $H_{2n+1}(x)$ already satisfies the Lagrange interpolating conditions

$$H_{2n+1}(x_i) = y_i, \quad i = 0, 1, \ldots, n,$$

for any polynomial $H_n(x)$.
Therefore, to determine $H_{2n+1}(x)$, we look for a polynomial $H_n(x)$ such that

$$H'_{2n+1}(x_i) = y'_i, \quad i = 0, 1, \ldots, n.$$

By differentiation of the formula (2.17), we have

$$H'_{2n+1}(x) = L'_n(x) + w'_{n+1}(x) H_n(x) + w_{n+1}(x) H'_n(x).$$

Hence

$$y'_i = L'_n(x_i) + w'_{n+1}(x_i) H_n(x_i),$$

$$H_n(x_i) = \frac{y_i' - L_n'(x_i)}{w_{n+1}'(x_i)}.$$

and, we get

$$H_n(x) = \sum_{i=0}^{n} \frac{y_i' - L_n'(x_i)}{w_n'(x_i)} l_i(x),$$

where $l_i(x)$, $i = 0, 1, ..., n$, are Lagrange polynomials. Finally, combining the above equalities, we obtain the following Hermite interpolating polynomial:

$$H_{2n+1}(x) = L_n(x) + w_{n+1}(x) \sum_{i=0}^{n} \frac{y_i' - L_n'(x_i)}{w_{n+1}'(x_i)} l_i(x). \quad (2.18)$$

2.6 Relationship Between a Function and its Hermite Interpolating Polynomial.

Let $H_m(x)$ be Hermite interpolating polynomial to a given function $f(x)$, *i.e.*

$$H_m^{(s_i-1)}(x_i) = f^{(s_i-1)}(x_i), \quad i = 0, 1, \ldots, n, \quad s_0 + s_1 + \cdots + s_n = m+1.$$

Then, the following theorem holds:

Theorem 2.3 *If the function $f(x)$ is $(m+1)$ times continuously differentiable on a closed interval $[a, b]$, then there exists a point $\xi_x \in (a, b)$ such that*

$$f(x) = H_m(x) + \frac{f^{(m+1)}(\xi_x)}{(m+1)!} \Omega_{m+1}(x), \quad x \in [a, b],$$

where

$$\Omega_{m+1}(x) = (x - x_0)^{s_0} (x - x_1)^{s_1} \cdots (x - x_n)^{s_n}.$$

Proof. Let us consider the auxiliary function

$$g(t) = f(t) - H_m(t) - K\Omega_{m+1}(t),$$

of the variable $t \in [a, b]$ for fixed $x \neq x_i$, $\quad i = 0, 1, \ldots, n$, where

$$K = \frac{f(x) - H_m(x)}{\Omega_{m+1}(x)}.$$

The function $g(t)$ has zeros at the points $x_0, x_1, \ldots, x_n$; of multiplicity $s_0, s_1, \ldots, s_n$, respectively. Also, $g(x) = 0$. Therefore, $g(t)$ has the total number of zeros in the interval $[a, b]$ equal

$$s_0 + s_1 + \cdots + s_n + 1 = m + 2.$$

By the Rolle's theorem, the first derivative $g'(t)$ possesses $m + 1$ zeros in the interval (a, b), the second derivative $g''(t)$ has m zeros in the interval (a, b), and finally, the derivative $g^{(m+1)}(t)$ has one zero in the interval (a, b). Therefore, there exists a point $\xi_x \in (a, b)$ such that

$$g^{(m+1)}(\xi_x) = 0.$$

On the other hand

$$g^{(m+1)}(\xi_x) = f^{(m+1)}(\xi_x) - K(m+1)!$$

and

$$f^{(m+1)}(\xi_x) - \frac{f(x) - H_m(x)}{\Omega_{m+1}(x)}(m+1)! = 0.$$

Hence, we obtain the formula

$$f(x) = H_m(x) + \frac{f^{(m+1)}(\xi_x)}{(m+1)!}\Omega_{m+1}(x), \quad x \in [a, b].$$

Clearly, from the theorem, the error of Hermite interpolation

$$f(x) - H_m(x) = \frac{f^{(m+1)}(\xi_x)}{(m+1)!}\Omega_{m+1}(x), \quad x \in [a, b].$$

satisfies the inequality

$$|f(x) - H_m(x)| \leq \frac{M^{(m+1)}}{(m+1)!}|\Omega_{m+1}(x)|, \quad x \in [a, b], \quad (2.19)$$

where

$$M^{(m+1)} = \max_{a \leq x \leq b} |f^{(m+1)}(x)|.$$

In the case when $s_0 = s_1 = \cdots = s_n = 2$, by (2.13), we arrive at the following error estimate

$$|f(x) - H_{2n+1}(x)| \leq \frac{M^{(2n+2)} n! n!}{16(2n+2)!} h^{2n+2}, \quad x \in [a, b], \quad (2.20)$$

Example 2.8 *Find Hermite interpolating polynomial $H_5(x)$ for the function $f(x) = e^x$ which satisfies the following conditions*

$$H_5(x_i) = e^{x_i}, \quad i = 0, 1, 2,$$

$$H_5'(x) = e^{x_i}, \quad i = 0, 1, 2,$$

where $x_0 = 0$, $x_1 = 1$, $x_2 = 2$.
Estimate the error of interpolation when $x \in [0, 2]$.

Solution. In order to find Hermite interpolating polynomial $H_5(x)$ for the function $f(x) = e^x$, we shall determine the Lagrange interpolating polynomial $P_2(x)$ spaned by the points 0, 1 and 2.
Thus, we have

$$
\begin{aligned}
L_2(x) &= \frac{(x-1)(x-2)}{(0-1)(0-2)} 1 \\
&+ \frac{x(x-2)}{(1-0)(1-2)} 2.7183 \\
&+ \frac{x(x-1)}{(2-0)(2-1)} 7.3891 = 1.47625 x^2 + 0.24205 x + 1,
\end{aligned}
$$

and $L_2'(x) = 2.9525 x + 0.24205$.
Next, we have

$$w_3(x) = x(x-1)(x-2) = x^3 - 3x^2 + 2x \text{ and } w_3'(x) = 3x^2 - 6x + 2.$$

Hence, by formula (2.18), we obtain the following Hermite polynomial

$$
\begin{aligned}
H_5(x) &= L_2(x) + w_3(x)[\frac{y_0' - L_2'(0)}{w_3'(0)}l_0(x) + \frac{y_1' - L_2'(1)}{w_3'(1)}l_1(x) \\
&+ \frac{y_2' - L_2'(2)}{w_3'(2)}l_2(x)] \\
&= 1.4765x^2 + 0.24205x + 1 \\
&+ (x^3 - 3x^2 + 2x)[\frac{1 - 0.242}{2}\frac{(x-1)(x-2)}{(0-1)(0-2)} \\
&+ \frac{2.7183 - 3.19455}{-1}\frac{x(x-2)}{(1-0)(1-2)} \\
&+ \frac{7.3891 - 6.14705}{2}\frac{x(x-1)}{(2-0)(2-1)}] \\
&= 0.0237209x^5 + 0.00240976x^4 + 0.2057x^3 \\
&+ 0.48647x^2 + x + 1.
\end{aligned}
$$

By (2.19), we get the following error estimate:

$$
|e^x - H_5(x)| \le \frac{e^{\xi x}}{6!}\max_{0 \le x \le 2} x^2(x-1)^2(x-2)^2 \le \frac{e^2}{6!}\frac{4}{27} \approx 0.0015.
$$

for all $x \in [0, 2]$.

We find the Hermite's interpolating polynomial $H_5(x)$ (*cf.* [2,3], [4],[5],[6], [12], [13]) by the following program:

```
data={{{0,{1,1}},{1,{Exp[1],Exp[1]}},
    {2,{Exp[2],Exp[2]}}};
H5=InterpolatingPolynomial[data,x];
N[Expand[H5],6]
```

2.7 Trigonometric Polynomials

Let $f(x)$ be a continuous and periodic function with the period $2L > 0$, so that

$$
f(x + 2L) = f(x) \quad \text{for all real } x.
$$

Below, we shall consider the trigonometric polynomial in the following form:

$$TR_n(x) = \frac{1}{2}a_0 + \sum_{k=1}^{n} a_k cos\frac{k\pi x}{L} + b_k sin\frac{k\pi x}{L},$$

where

$$a_k = \frac{1}{L} \int_{-L}^{L} f(x)cos\frac{k\pi x}{L}dx,$$

$$b_k = \frac{1}{L} \int_{-L}^{L} f(x)sin\frac{k\pi x}{L}dx.$$

Now, let us consider the Riemann's sums of the above integrals

$$a_k \approx \frac{1}{L} \sum_{j=-n}^{n} f(x_j)cos\frac{k\pi x_j}{L}\Delta x = \alpha_k,$$

$$b_k \approx \frac{1}{L} \sum_{j=-n}^{n} f(x_j)sin\frac{k\pi x_j}{L}\Delta x = \beta_k,$$

(2.21)

where

$$\Delta x = \frac{2L}{2n+1}, \quad x_j = \frac{2L}{2n+1}j$$

for $j = -n, -n+1, -n+2, \ldots, 0, 1, \ldots, n$.
The following theorem holds:

Theorem 2.4 *The interpolating trigonometric polynomial* $TR_n(x)$ *for the function* $f(x)$ *at the point* $x_j = \Delta x_j$, $j = -n, -n+1, -n+2, \ldots, 0, 1, \ldots, n$ *which satisfies the conditions*

$$TR_n(x_j) = f(x_j), \quad j = -n, -n+1, -n+2, \ldots, 0, 1, \ldots, n,$$

is given by the formula

$$TR_n(x) = \frac{1}{2}\alpha_0 + \sum_{k=1}^{n} \alpha_k cos\frac{k\pi x}{L} + \beta_k sin\frac{k\pi x}{L},$$

where the coefficients α_k and β_k are determined by formula (2.21).

Proof. We note that

$$
\begin{aligned}
TR_n(x_j) \; = \; & \frac{1}{2L} \sum_{s=-n}^{n} f(x_s)\Delta x \\
& + \sum_{k=1}^{n} \frac{1}{L} \sum_{s=-n}^{n} f(x_s) \cos\frac{k\pi x_s}{L} \cos\frac{k\pi x_j}{L} \Delta x \\
& + \sum_{k=1}^{n} \frac{1}{L} \sum_{s=-n}^{n} f(x_s) \sin\frac{k\pi x_s}{L} \sin\frac{k\pi x_j}{L} \Delta x \\
= \; & \frac{1}{L} \sum_{-n}^{n} f(x_s)\Delta x [\frac{1}{2} + \sum_{k=1}^{n} \cos\frac{k\pi x_s}{L} \cos\frac{k\pi x_j}{L} \\
& + \sin\frac{k\pi x_s}{L} \sin\frac{k\pi x_j}{L}] \\
= \; & \frac{1}{L} \sum_{s=-n}^{n} f(x_s)\Delta x [\frac{1}{2} + \sum_{k=1}^{n} \cos(x_s - x_j)\frac{k\pi}{L}].
\end{aligned}
$$

Hence, by the following trigonometric identity

$$
\frac{1}{2} + \sum_{k=1}^{n} \cos\mu k = \begin{cases} \frac{1}{2} + n & \text{if } \mu = 0, \\ 0 & \text{if } \mu \neq 0, \end{cases}
$$

we obtain

$$
TR_n(x_j) = \frac{1}{L} f(x_j)\Delta x (\frac{1}{2} + n) = f(x_j),
$$

for $j = -n, -n+1, -n+2, \ldots, 0, 1, \ldots, n$.

Thus, $TR_n(x)$ is the interpolating trigonometric polynomial to the periodic function $f(x)$ in the interval $[-L, L]$.

The conditions for convergence of the sequence $TR_n(x)$, $n = 1, 2, \ldots$; to the function $f(x)$ are presented in the following theorem:

Theorem 2.5 *If $f(x)$ is a continuous function in the interval $[-L, L]$ except at a finite number of points $x_0, x_1, \ldots, x_m$ at which $f(x)$ has both left and right sides limits i.e.,*

$$
\lim_{x \to x_k+} f(x) = f(x_k+) \quad \text{and} \quad \lim_{x \to x_k-} f(x) = f(x_k-).
$$

then the sequence $TR_n(x), \quad n = 1, 2, \ldots;$ *is convergent and*

$$\lim_{n \to \infty} TR_n(x) = \frac{1}{2}[f(x+) + f(x-)]$$

for all $x \in [-L, L]$.

Example 2.9 *Find an interpolating trigonometric polynomial* $TR_2(x)$ *for the following function*

$$f(x) = x - r \quad \text{for} \quad r - 1 \le x < r + 1, \quad r = 0, \pm 2, \pm 4, \ldots;$$

Solution. Let us note that $f(x)$ is a periodic function with the period $2L = 2$. Therefore, we may consider $f(x)$ only for $x \in [-1, 1]$ setting $r = 0$. Then, we have

$$n = 2, \quad 2L = 2, \quad \Delta x = 0.4, \quad x_j = 0.4j, \quad j = -2, -1, 0, 1, 2.$$

Thus

$$TR_2(x) = \frac{1}{2}\alpha_0 + \sum_{j=-2}^{2} \alpha_j \cos j\pi x + \beta_j \sin j\pi x,$$

where

$$\alpha_0 = \sum_{j=-2}^{2} 0.16 \ j = 0,$$

$$\alpha_1 = \sum_{j=-2}^{2} 0.16 \ j \ cos \ 0.4\pi j = 0,$$

$$\alpha_2 = \sum_{j=-2}^{2} 0.16 \ j \ cos \ 2 * 0.4\pi \ j = 0,$$

and

$$\beta_1 = \sum_{j=-2}^{2} 0.16 \ j \ sin \ 0.4 \ \pi \ j = 0.6805206,$$

$$\beta_2 = \sum_{j=-2}^{2} 0.16 \ j \ sin \ 0.8 \ j \ \pi = -.42058489.$$

Hence, the interpolating trigonometric polynomial is

$$TR_2(x) = 0.6805206 \ sin \ \pi x \ - \ 0.42058489 \ sin \ 2\pi x.$$

2.8 Relationship between Trigonometric Polynomials and Fourier's Transformation

Let us assume that $f(x)$ is a continuous function for all real x and there exists the improper integral

$$\int_{-\infty}^{\infty} |f(x)|dx \ < \infty.$$

The trigonometric polynomial $TR_n(x)$ for the function $f(x)$ is:

$$
\begin{aligned}
TR_n(x) &= \frac{a_0}{2} + \sum_{k=1}^{n} a_k cos\frac{k\pi x}{L} + b_k sin\frac{k\pi x}{L} \\
&= \frac{a_0}{2} + \sum_{k=1}^{n} a_k \frac{e^{i\frac{k\pi x}{L}} + e^{-i\frac{k\pi x}{L}}}{2} \\
&+ \sum_{k=1}^{n} b_k \frac{e^{i\frac{k\pi x}{L}} - e^{-i\frac{k\pi x}{L}}}{2i} = \sum_{k=-n}^{n} A_k e^{i\frac{k\pi x}{L}},
\end{aligned}
$$

where $-L \leq x \leq L$, and the coefficients

$$A_k = \frac{1}{2L} \int_{-L}^{L} f(t)e^{-i\frac{k\pi x}{L}} dt.$$

Hence

$$TR_n(x) = \sum_{k=-n}^{n} \left(\frac{1}{2L} \int_{-L}^{L} f(t)e^{-i\frac{k\pi t}{L}} dt\right)e^{i\frac{k\pi x}{L}}$$

Let us put

$$s_k = \frac{k\pi}{L}, \qquad \Delta_k = s_{k+1} - s_k = \frac{\pi}{L}.$$

Then, we have

$$TR_n(x) = \frac{1}{2L} \int_{-L}^{L} \left(\sum_{k=-n}^{n} f(t)e^{-is_k t} e^{is_k x}\right)dt.$$

Now, we note that

$$TR_n(x) \rightarrow f(x) \ \text{ and } \ \sum_{k=-n}^{n} e^{i(x-t)s_k} \Delta s_k \rightarrow \int_{-\infty}^{\infty} e^{i(x-t)s} ds,$$

when $n \to \infty$ and $L \to \infty$.
Hence

$$f(x) = \frac{1}{2L} \int_{-\infty}^{\infty} \left(\int_{-\infty}^{\infty} f(t) e^{-ist} dt \right) e^{isx} ds. \qquad (2.22)$$

Hence, we arrive at the Fourier transform of $f(x)$

$$F(s) = \frac{1}{\sqrt{2L}} \int_{-\infty}^{\infty} f(t) e^{-ist} dt,$$

and, by (2.22), we get the inverse Fourier transform

$$f(x) = \frac{1}{\sqrt{2L}} \int_{-\infty}^{\infty} F(s) e^{ist} ds.$$

2.9   Exercises

Question 2.1 *The function $f(x) = \sqrt{6+x}$ is tabulated below .*

x	-2	-1	1	2
$f(x)$	2	2.2336	2.6446	2.8284

.

1. (a) *Find the Newton's interpolating polynomial and reduce it to the natural form.*

    (b) *Calculate the approximate value of $f(1.5)$ and give an estimate of the error of interpolation for every $x \in [-2, 2]$.*

Question 2.2 *The function $f(x) = \ln(6+x)$ is tabulated below .*

x	-2	-1	1	2
$f(x)$	1.3863	1.6094	1.9459	2.0794

.

1. (a) *Find the Lagrange's interpolating polynomial and reduce it to the natural form.*

(b) *Calculate the approximate value of $f(1.5)$ and give an estimate of the error of interpolation for every $x \in [-2, 2]$.*

Question 2.3 *Consider the following function*

$$f(x) = \frac{1}{2 + x}, \qquad 0 \le x \le 1,$$

1. (a) *Find the Hermite interpolating polynomial $H_3(x)$ that satisfies the following interpolating conditions*

$$H_3(i) = f(i), \qquad H_3'(i) = f'(i), \quad i = 0, 1.$$

(b) *Estimate the error of interpolation.*

Question 2.4 *Find an optimal interpolating polynomial of degree at most two for the following functions*

(i) $f(x) = sin \frac{\pi}{2}x, \qquad -1 \le x \le 1,$

(ii) $f(x) = ln(1 + x), \qquad 0 \le x \le 2.$

Estimate the error of interpolation.

Question 2.5 *Assume that $f(x)$ is a given function at the knots*

$$a \le x_0 < x_1 < \cdots < x_k \le b.$$

Show that the divided difference

$$[x_0 x_1 ... x_k]$$

of order k can be written in the following form

$$[x_0 x_1 ... x_k] = \frac{f(x_0)}{(x_0 - x_1)(x_0 - x_2) \cdots (x_0 - x_k)}$$

$$+ \frac{f(x_1)}{(x_1 - x_0)(x_1 - x_2) \cdots (x_1 - x_k)}$$

$$\cdots \quad \ldots\ldots\ldots\ldots\ldots\ldots\ldots\ldots\ldots\ldots\ldots\ldots\ldots\ldots$$

$$+ \frac{f(x_k)}{(x_k - x_0)(x_k - x_1) \cdots (x_k - x_{k-1})}$$

Hint: Use the principle of mathematical induction.

Question 2.6 *Let $f(x)$ be a function $(n+1)$ times contin-uously differentiable in interval $[a, b]$.*

1. (a) *Use the following formula*

$$f(x) = L_n(x) + (x - x_0)(x - x_1) \cdots (x - x_n)[xx_0x_1x_2 \cdots x_n]$$

   large to show that there exits a point $\xi \in (a, b)$ for which

$$[xx_0x_1 \cdots x_n] = \frac{f^{(n+1)}(\xi)}{(n+1)!}$$

   (b) Find $[xx_0x_1x_2 \cdots x_n]$ for $f(x) = x^{n+1}$.

Question 2.7 *Let $f(x) = x^n$. Show that*

$$[x_0, x_1...x_k] = \begin{cases} \displaystyle\sum_{\alpha_0+\alpha_1+\cdots+\alpha_k=n-k} x_0^{\alpha_0} x_1^{\alpha_1} ... x_k^{\alpha_k}, & \text{if } n > k, \\ 1 & \text{if } n = k, \\ 0 & \text{if } n < k. \end{cases}$$

Hint: Use the principle of mathematical induction.

Question 2.8 *Find an interpolating trigonometric polyno-mial $TR_3(x)$ to the following function*

$$f(x) = x - r, \quad \text{for } r - 1 \le x < r + 1, \quad r = 0, \pm 2, \pm 4, \ldots;$$

Send Orders for Reprints to reprints@benthamscience.net

Lecture Notes in Numerical Analysis with Mathematica, 2014, 63-102   **63**

CHAPTER 3

Polynomial Splines

Abstract

In the chapter, the space $S_m(\Delta, m-1)$ of piecewise polynomial splines of degree m and differentiable up to the order $m-1$ is introduced. In particular, the space $S_1(\Delta, 0)$ of piecewise linear splines and the space $S_3(\Delta, 2)$ of cubic splines are determined . Theorems on interpolation by the splines are stated and proved. The space $S_{11}(\Delta, 0, 0)$ of belinear splines and the space $S_{33}(\Delta, 2, 2)$ of be-cubic splines in two variables defined on rectangular grids are presented. On triangular grids the spaces $\Pi_1(\Delta)$ and $\Pi_3(\Delta)$ are considered. Mathematica modules have been designed for solving problems associated with application of splines. The chapter ends with a set of questions.

Keywords: Polynomial splines, linear splines, cubic splines.

3.1   Space $S_m(\Delta, k)$.

Polynomial splines of the class $S_m(\Delta, k)$ are successfully applied in the theory of approximation of functions as well as in solving of problems which arise in the fields of differential equations and engineering.

In order to introduce the definition of polynomial splines of degree m, let us first define normal partition Δ of the interval $[a, b]$.
A partition

$$\Delta \; : \quad a = x_0 < x_1 < \cdots < x_N = b,$$

is normal if there exits constant σ such that

$$\frac{\max_{0 \leq i \leq N-1}(x_{i+1} - x_i)}{\min_{0 \leq i \leq N-1}(x_{i+1} - x_i)} = \sigma_N,$$

and $\sigma_N \leq \sigma$ for all natural N.

Definition 3.1 *A function $s(x)$ is said to be a polynomial spline of degree m if the following conditions are satisfied:*

- *$s(x)$ is a polynomial of degree at most m on each subinterval $[x_i, x_{i+1}]$,*
  $i = 0, 1, \ldots, N - 1$.

- *$s(x)$ and its first $m-1$ derivatives are continuous functions on the interval $[a, b]$.*

The class of all polynomial splines of degree m spanned over the partition Δ shall be denoted by the symbol $S_m(\Delta, m-1)$.

The Basis. Now, we shall determine a basis of the space $S_m(\Delta, m - 1)$. Let us consider the following auxiliary function:

$$(x - t)_+^m = \begin{cases} (x - t)^m & \text{if} \;\; x \leq t, \\ 0 & \text{if} \;\; x > t. \end{cases}$$

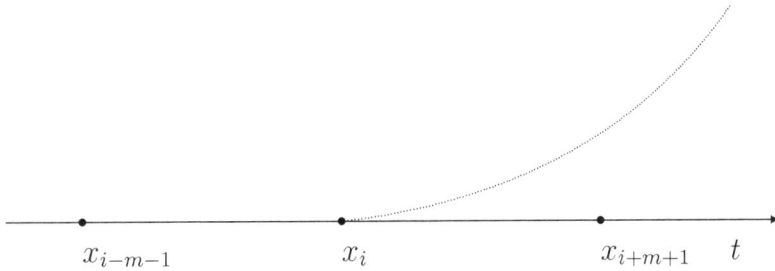

$$x_{i-m-1} \qquad\qquad x_i \qquad\qquad x_{i+m+1} \qquad t$$

Fig. 3.1. Auxiliary function $(x_i - t)_+^m$

The finite difference of order $m+1$ of the auxiliary function is

$$\Delta^{m+1}(x_i - t)_+^m = (-1)^{m-1} \sum_{\nu=0}^{m+1} (-1)^\nu \binom{m+1}{\nu} (x_{i+\nu} - t)_+^m.$$

Example 3.1 *We compute the differences*

$$m = 1, \quad \Delta^2 (x_i - t)_+ = (x_i - t)_+ - 2(x_{i+1} - t)_+$$
$$+ \quad (x_{i+2} - t)_+$$

$$m = 2, \quad \Delta^3 (x_i - t)_+^2 = -(x_i - t)_+^2 + 3(x_{i+1} - t)_+^2$$
$$- \quad 3(x_{i+2} - t)_+^2 - (x_{i+3} - t)_+^2$$

$$m = 3, \quad \Delta^4 (x_i - t)_+^3 = (x_i - t)_+^3 - 3(x_{i+1} - t)_+^3$$
$$+ \quad 6(x_{i+2} - t)_+^3 - 3(x_{i+3} - t)_+^3$$
$$+ \quad (x_{i+4} - t)_+^3$$

Assuming that Δ is a uniform partition of interval $[a, b]$, so that

$$x_i = a + ih, \qquad i = 0, 1, ..., N; \qquad h = \frac{b - a}{N}.$$

we find the difference

$$\Delta^{m+1}(x_i - t)_+^m = 0 \quad \text{for} \quad t \geq x_i.$$

Indeed, the function

$$(x_{i+\nu} - t)_+^m = (x_{i+\nu} - t)^m, \quad \text{for } t \geq x_{i+\nu}, \quad \nu = 0, 1, ..., m+1$$

is the polynomial of degree m, therefore its difference of order $m + 1$ is equal to zero.

Also, we note that

$$\Delta^{m+1}(x_i - t)_+^m = 0 \quad \text{for} \quad t \leq x_i,$$

Because

$$(x_{i+\nu} - t)_+^{m+1} = 0, \quad t \leq x_i$$

for $\nu = 0, 1, .., m + 1$.

Hence, we have

$$\Delta^{m+1}(x_i - t)_+^m = \qquad\qquad (3.1)$$

$$\begin{cases} \displaystyle\sum_{\nu=0}^{m+1} (-1)^{\nu+m+1} \binom{m+1}{\nu}(x_{i+\nu} - t)_+^m, & x_{i+q-1} < t < x_{i+q}, \\ & q = 1, 2, ..., m+1, \\ \\ 0, & t \leq x_i \text{ or } t > x_{i+m+1} \end{cases}$$

It is clear, by (3.1), that the functions

$$K_m(x_i, t) = \Delta^{m+1}(x_i - t)_+^m, \qquad i = 0, 1, ..., N, \qquad (3.2)$$

are polynomials of degree m on each subinterval $[x_i, x_{i+1}]$, and they are $m - 1$ times continuously differentiable in the interval $[a, b]$, i.e., $K_m(x_i, t) \in C^{m-1}[a, b]$. Therefore, each $K_m(x_i, t)$ is a polynomial spline of the class $S_m(\Delta, m - 1)$. Obviously, $K_m(x_i, t)$ can be considered on the whole real line with infinite number of knots x_i, $i = 0, \pm1, \pm2, ...$; However, only $N+m$ of them are not identically equal to zero on the interval $[a, b]$. These non-zero and linearly independent splines are:

$$K_m(x_{-m}, t), \ K_m(x_{-m+1}, t), \ K_m(x_{-m+2}, t), \ \cdots, K_m(x_{N-1}, t).$$

Example 3.2 *Let $m = 1$, then, by (3.2), (see* **Fig. 3.1**)*, we obtain*

$$K_1(i,t) = \begin{cases} 0, & t \le i, \quad \text{or} \quad t \ge i+2, \\ -i+t, & i \le t \le i+1, \\ i-t+2, & i+1 \le t \le i+2. \end{cases}$$

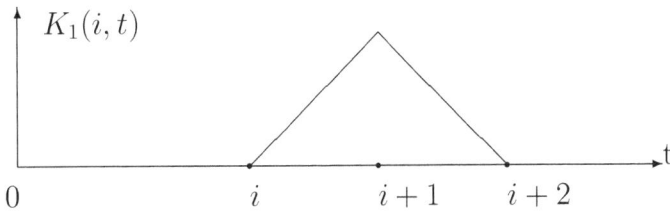

Fig 3.2. Linear spline $K_1(i,t)$

For the uniform partion of the interval $[a, b]$, we shall consider normalized splines

$$\frac{1}{h^m} K_m(x_{i-m+1}, x), \quad i = 1, 2, ..., N+3, \quad a \le x \le b,$$

as a basis of the space $S_m(\Delta, m-1)$ of the dimension

$$dim(S_m(\Delta, m-1)) = N + m.$$

3.2　Properties of Splines.

Minimum property. Below, we shall consider the space $S_m(\Delta, m-1)$, when $m = 2q + 1$ is an odd positive integer. Splines from this space minimize the following functional

$$F(g) = \int_a^b [g^{(q+1)}(x)]^2 dx, \quad g \in C^{(q+1)}[a, b].$$

Precisely, let us consider the following variational problem (*cf.* [1,2,7,9, 22,23,25,26]):

Variational Problem. *Find a function $s \in C^{(q+1)}[a, b]$ at* *which the functional $F(g)$ attains its minimum under the interpolation conditions:*

$$g(x_i) = f(x_i), \quad i = 0, 1, ..., N,$$

and either

$$g^{(j)}(a) = g^{(j)}(b) = 0, \quad j = q + 1, q + 2, ..., 2q, \quad (3.3)$$

or

$$g^{(j)}(a) = g^{(j)}(b) = 0, \quad j = 1, 2, ..., q,$$

for a given function $f(x)$, $x \in [a, b]$.
[1] The following theorem holds:

Theorem 3.1 *There exists a unique spline $s \in S_m(\Delta, m - 1)$ which satisfies the interpolation conditions (3.3).*

Proof. Every spline $s \in S_m(\Delta, m - 1)$ can be written as the following linear combination:

$$s(x) = a_{-m}K_m(x_{-m}, x) + a_{-m+1}K_{-m+1}(x_{m+1}, x) + \cdots$$

$$+ \quad a_{N-1}K_m(x_{N-1}, x).$$

By the interpolation conditions, $N + m$ coefficients

$$a_{-m}, a_{-m+1}, ..., a_{N-1}$$

have to satisfy the following system of $N + m$ linear equations:

$$s(x_i) = f(x_i), \quad i = 0, 1, ..., N$$

$$s^{(j)}(a) = s^{(j)}(b) = 0, \quad j = q + 1, q + 2, ..., 2q, \quad m = 2q + 1.$$

Clearly, there exists a unique interpolating spline

$$s \in S_m(\Delta, m - 1)$$

if the corresponding system of homogeneous equations possesses only trivial solution. To show this, we assume that

---

[1] The conditions put on the derivatives can be replaced by periodicity conditions.

$f(x_i) = 0$, for $i = 0, 1, ..., N$.
Then, we have

$$F(s) = \int_a^b [s^{(q+1)}(x)]^2 dx$$

$$= -\int_a^b s^{(q)}(x) s^{(q+2)}(x) dx$$

$$\cdots\cdots\cdots\cdots\cdots\cdots\cdots\cdots\cdots\cdots$$

$$= (-1)^q \int_a^b s'(x) s^{(2q+1)}(x) dx$$

$$(-1)^q \sum_{i=0}^{N-1} \{ s(x_{i+1}) s^{(2k+1)}(x_{i+1}) - s(x_i) s^{(2k+1)}(x_i)$$

$$- \int_{x_i}^{x_{i+1}} s(x) s^{(2q+2)}(x) dx \}.$$

Hence, $F(s) = 0$, since $s(x_i) = 0$, for $i = 0, 1, ..., N$ and $s^{(2q+2)}(x) = 0$, for $x_i \leq x \leq x_{i+1}$. Thus, $s^{(q+1)}(x) = 0$ in the interval $[a, b]$. Therefore, $s(x)$ is a polynomial of degree at most q which has at least $N + 1$, $(q \leq N)$ roots in the interval $[a, b]$. Then, $s(x) = 0$, for $x \in [a, b]$ and $a_{-m} = a_{-m+1} = ... = a_{N-1} = 0$. This means that $s(x)$ is a unique interpolating spline in $S_m(\Delta, m - 1)$.
The following theorem holds:

Theorem 3.2 *There exists a unique solution $s \in S_m(\Delta, m-1)$ of the variational problem.*

Proof. Let $s \in S_m(\Delta, m - 1)$ be the unique interpolating spline which satisfies the conditions (3.3). Then, we note that

$$\int_a^b [g^{(q+1)}]^2 dx = \int_a^b [s^{(q+1)}]^2 dx + \int_a^b [g^{(q+1)} - s^{(q+1)}(x)]^2 dx$$

$$+ 2 \int_a^b s^{(q+1)}(x)(g^{(q+1)}(x) - s^{(q+1)}(x)) dx.$$

Integrating by parts under the conditions (3.3), we find

$$\int_a^b s^{(q+1)}(x)[g^{(q+1)}(x) - s^{(q+1)}(x)]dx$$

$$= \sum_{i=0}^{N-1} \int_{x_i}^{x_{i+1}} s^{(q+1)}(x)[g^{(q+1)}(x) - s^{(q+1)}(x)]dx =$$

$$\cdots\cdots\cdots\cdots\cdots\cdots\cdots\cdots\cdots\cdots\cdots\cdots\cdots\cdots\cdots$$

$$= (-1)^q \sum_{i=0}^{N-1} \int_{x_i}^{x_{i+1}} s^{(2q+1)}(x)[g'(x) - s'(x)]dx$$

$$= (-1)^{q+1} \sum_{i=0}^{N-1} \int_{x_i}^{x_{i+1}} s^{(2q+2)}(x)[g(x) - s(x)]dx.$$

Recalling the equality

$$s^{(2k+2)}(x) = 0 \quad \text{for} \quad x_i \leq x \leq x_{i+1}, \quad i = 0, 1, ..., N,$$

we find

$$\int_a^b s^{(q+1)}(x)(g^{(q+1)}(x) - s^{(q+1)}(x))dx = 0.$$

and

$$F(g) = \int_a^b [g^{(q+1)}(x)]^2 = \int_a^b [s^{(q+1)}(x)]^2 dx$$

$$- \int_a^b [g^{(q+1)}(x) - s^{(q+1)}(x)]^2 dx$$

Hence, the functional $F(g)$ attains minimum at $g(x) = s(x)$, $x \in [a, b]$.

In order to prove that there is a unique spline which minimizes the functional $F(g)$, let us assume for contrary that there are at least two such splines $s_1(x)$ and $s_2(x)$. Then, the difference $s(x) = s_1(x) - s_2(x)$ is the spline which satisfies homogeneous interpolation conditions. Therefore, $s(x) \equiv 0$, $x \in [a, b]$.

Example 3.3 *Let us consider the space* $S_3(\Delta, 2)$ *of cubic splines. Then, the interpolation conditions (3.3) take the following form:*

$$g(x_i) = f(x_i), \quad i = 0, 1, ..., N, \quad g''(a) = g''(b) = 0. \quad (3.4)$$

By theorems (4.2) and (4.3), there exists a unique cubic spline $s \in S_3(\Delta, 2)$ *at which the functional*

$$F(g) = \int_a^b [g''(x)]^2 dx, \quad g \in C^{(2)}[a, b],$$

attains minimum under the interpolation conditions (3.4), i.e.

$$F(s) \leq \int_a^b [g''(x)]^2 dx, \quad for \ all \quad g \in C^{(2)}[a, b].$$

3.3 Examples of Polynomial Splines

Space $S_1(\Delta, 0)$ **of Piecewise Linear Splines.** Elements of the space $S_1(\Delta, 0)$ (**see Fig. 3.2**) are piecewise linear splines of the following form:

$$s(x) = a_0 \psi_0(x) + a_1 \psi_1(x) + \cdots + a_N \psi_N(x),$$

where the basis splines

$$\psi_i(x) = \begin{cases} \dfrac{x - x_{i-1}}{x_i - x_{i-1}} & if \quad x_{i-1} \leq x < x_i \\ \dfrac{x_{i+1} - x}{x_{i+1} - x_i} & if \quad x_i \leq x < x_{i+1} \\ 0 & if x < x_{i-1} \ \ or \ \ x \geq x_{i+1} \end{cases} \quad (3.5)$$

For the uniform distribution of the points $x_i = ih$, $i = 0, \pm 1, \pm 2 \ldots$, the piecewise linear splines $\psi_i(x)$ are given by the following formulae:

$$\psi_i(x) = \frac{1}{h} \begin{cases} x - x_{i-1} & if \quad x_{i-1} \leq x < x_i \\ x_{i+1} - x & if \quad x_i \leq x < x_{i+1} \\ 0 & if x < x_{i-1} \ \ or \ \ x \geq x_{i+1} \end{cases} \quad (3.6)$$

Let us note that there are $N + 1$ piecewise linear splines not identically equal to zero on the interval $[a, b]$. Thus, the space of piecewise linear splines

$$S_1(\Delta, 0) = span\{\psi_0, \psi_1, \ldots, \psi_N\}$$

has dimension $N + 1$.

Now, we observe that every piecewise linear spline $s(x)$ can be written as the following linear combination:

$$s(x) = a_0\psi(x) + a_1\psi_1(x) + \cdots + a_N\psi_N(x).$$

In particular, we find the Lagrange's interpolating piecewise linear spline

$$s(x) = f(x_0)\psi_0(x) + f(x_1)\psi_1(x) + f(x_2)\psi_2(x) + \cdots + f(x_N)\psi_N(x),$$

to a given function $f(x)$ which satisfies the following conditions of interpolation:

$$s(x_i) = f(x_i), \quad i = 0, 1, \ldots, N.$$

The following theorem holds:

Theorem 3.3 *If f is a function twice continuously differentiable on the interval $[a, b]$, then the error of interpolation $f(x) - s(x)$ satisfies the inequality:*

$$\mid f(x) - s(x) \mid \leq \frac{h^2}{8} M^{(2)}, \quad x \in [a, b].$$

Proof. Let $x_i \leq x \leq x_{i+1}$. Then, we have

$$f(x) - s(x) = \frac{f''(\xi_x)}{2!}(x - x_i)(x - x_{i+1}).$$

Since

$$\mid (x - x_i)(x - x_{i+1}) \mid \leq \frac{h^2}{4},$$

we get

$$\mid f(x) - s(x) \mid \leq \frac{h^2}{8} M^{(2)},$$

where

$$M^{(2)} = \max_{a \leq x \leq b} \mid f''(x) \mid.$$

Example 3.4 *Approximate the function*

$$f(x) = \sqrt{1+x}, \quad 0 \le x \le 2$$

by a piecewise linear spline with accuracy $\epsilon = 0,01$.

Solution. We shall start by determining h. The error of interpolation of a function $f(x)$ by a piecewise linear spline $s(x)$ satisfies the following inequality:

$$| f(x) - s(x) | \le \frac{h^2}{8} M^{(2)}, \quad a \le x \le b.$$

Since

$$M^{(2)} = \max_{a \le x \le b} | f''(x) | = \max_{a \le x \le b} \frac{1}{4\sqrt{(1+x)^3}} = \frac{1}{4},$$

we may choose h such that

$$\frac{h^2}{8} M^{(2)} = \frac{h^2}{32} \le \epsilon = 0.01.$$

Hence, we find $h = 0.5$.

For $h = 0.5$, $N = 4$, $x_0 = 0$, $x_1 = 0.5$, $x_2 = 1; x_3 = 1.5$, and $x_4 = 2$, the piecewise linear spline is given by the following formula

$$
\begin{aligned}
s(x) = \ & (2x(\sqrt{1.5} - 1) + 1)\theta_0(x) \\
& + (2x(\sqrt{2} - \sqrt{1.5}) + 2\sqrt{1.5} - \sqrt{2})\theta_1(x) \\
& + (2x(\sqrt{2.5} - \sqrt{2}) + 3\sqrt{2} - 2\sqrt{2.5})\theta_2(x) \\
& + (2x(\sqrt{3} - \sqrt{2.5}) + 4\sqrt{2.5} - 3\sqrt{3})\theta_3 = \\
& = \begin{cases}
0.4495x + 1 & \text{if } 0 \le x \le 0.5, \\
0.3789x + 1.0353 & \text{if } 0.5 \le x \le 1, \\
0.3339x + 1.0804 & \text{if } 1 \le x \le 1.5, \\
0.3018x + 1.1284 & \text{if } 1.5 \le x \le 2.
\end{cases}
\end{aligned}
$$

Now, let us solve this example using the following *Mathematica* module

Program 3.1 *Mathematica module that finds linear spline for a given data table.*

```
linearSpline[f_,a_,b_,n_,xstep_]:=Module[{h,sol},
 h=(b-a)/n;
 onex[x_]:=Module[{xr,r},
 xr=Table[a+r*h,{r,0,n+1}];
 k=Floor[(x-a)/h+1];
 N[f[xr[[k]]]+(f[xr[[k+1]]]
   -f[xr[[k]]])*(x-xr[[k]])/h]
  ];

 Print["   Linear spline approximating f(x) "];
 Print[" ----------------------"];
 linear=Table[{N[t],onex[t],N[f[t]]},
               {t,a,b,xstep}];
 TableForm[PrependTo[linear,{" x ",
             " linear ",f[x]}]]
  ]
```

By executing the following instructions

```
        f[x_]:=Sqrt[1+x];
        linearSpline[f,0,2,4,0.2];
```

we obtain the following table:

```
   linearSpline[f,0,2,4,0.2]
   Linear spline approximating f(x)
   ------------------------------------
   Out[3]/TableForm=
   x         linear      Sqrt[1+x]

   0         1.          1.
   0.2       1.0899      1.09545
```

0.4	1.1798	1.18322
0.6	1.26264	1.26491
0.8	1.33843	1.34154
1.	1.41421	1.41421
1.2	1.48098	1.48324
1.4	1.54775	1.54919
1.6	1.61132	1.61245
1.8	1.67169	1.67322
2.	1.73205	1.73205

Space $S_3(\Delta, 2)$ of Cubic Splines. As we know, a function f can be approximated by a piecewise linear spline with the accuracy $O(h^2)$. More accurate approximation of a smooth function can be found in the space $S_3(\Delta, 2)$ of cubic splines. To determine a base of the space $S_3(\Delta, 2)$, we start with the auxiliary function

$$K_3(x_i, t) = \Delta^4(x_i - t)_+^3, \quad x_i = ih, \quad i = 0. \pm 1 \pm 2, \ldots,$$

where

$$(x_i - t)_+^3 = \begin{cases} (x_i - t)^3 & \text{if } x \leq t, \\ 0 & \text{if } x > t. \end{cases}$$

We can consider the following cubic splines as the basis of the space $S_3(\Delta, 2)$ when the points x_i, $i = 0, \pm, 1, \pm 2, \ldots$, are uniformly distributed.

$$B_i(x) = \frac{1}{h^3} K_3(x_{i-2}, x), \quad i = -1, 0, 1, 2, \ldots, N + 1.$$

The explicit form of the cubic splines $B_i(x)$, $i = 0, \pm 1, \pm 2, ...,$ is given below:

$$
B_i(x) = \frac{1}{h^3}
\begin{cases}
0 \\
\text{if } x \le x_{i-2} \\[4pt]
(x - x_{i-2})^3 \\
\text{if } x_{i-2} \le x \le x_{i-1}, \\[4pt]
h^3 + 3h^2(x - x_{i-1}) + 3h(x - x_{i-1})^2 - 3(x - x_{i-1})^3 \\
\text{if } x_{i-1} \le x \le x_i, \\[4pt]
h^3 + 3h^2(x_{i+1} - x) + 3h(x_{i+1} - x)^2 - 3(x_{i+1} - x)^3 \\
\text{if } x_i \le x \le x_{i+1}, \\[4pt]
(x_{i+2} - x)^3 \\
\text{if } x_{i+1} \le x \le x_{i+2}, \\[4pt]
0 \\
\text{if } x \ge x_{i+2},
\end{cases}
\tag{3.7}
$$

for $i = 0, \pm 1, \pm 2, \ldots.$　$x \in (-\infty, \infty)$.

Now, let us note that the splines $B_i(x) = 0$ for $x \le x_{i-2}$ or $x \ge x_{i+2}$, so that the only non-zero cubic splines in the interval $[a, b]$ are $B_i(x)$ for $i = -1, 0, 1, \ldots, N+1, N+2,$. Therefore, any cubic spline $s(x)$ can be represented on the interval $[a, b]$ as a linear combination of cubic splines $B_i(x)$, $i = -1, 0, 1, ..., N+1$ i.e.,

$$s(x) = a_{-1}B_{-1}(x) + a_0 B_0(x) + \cdots + a_N B_N(x) + a_{N+1}B_{N+1}(x)$$

for $x \in [a, b]$.

x	x_{i-2}	x_{i-1}	x_i	x_{i+1}	x_{i+2}
$B_i(x)$	0	1	4	1	0
$B_i'(x)$	0	$\dfrac{3}{h}$	0	$-\dfrac{3}{h}$	0
$B_i''(x)$	0	$\dfrac{6}{h^2}$	$-\dfrac{12}{h^2}$	$\dfrac{6}{h^2}$	0

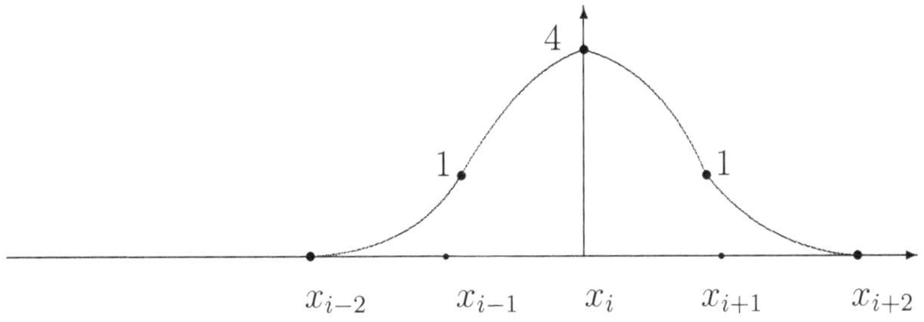

Fig. 3.3 $B_i(x)$

It can be proved that the cubic splines $B_{-1}(x), B_0(x), \ldots, B_{N+1}(x)$ create the complete basis of the space $S_3(\Delta, 2)$, (**see Fig. 3.3**)) so that

$$S_3(\Delta, 2) = span\{B_{-1}, B_0, \ldots, B_{N+1}\}.$$

In order to find an interpolating cubic spline $s(x)$ in the space $S_3(\Delta, 2)$, we can use the following conditions:

$$s'(x_0) = f'(x_0),$$
$$s(x_i) = f(x_i), \quad i = 0, 1, \ldots, N,$$
$$s'(x_N) = f'(x_N),$$

where $f(x)$ is a given function on the interval $[a, b]$. There exists only one interpolating cubic spline $s(x)$ which satisfies the above conditions. From these conditions, we get the following linear system of equations

$$
\begin{aligned}
s'(x_0) &= a_{-1}B'_{-1}(x_0) + a_0 B'_0(x_0) + \cdots + a_{N+1}B'_{N+1}(x_0) = f'(x_0) \\
s(x_0) &= a_{-1}B_{-1}(x_0) + a_0 B_0(x_0) + \cdots + a_{N+1}B_{N+1}(x_0) = f(x_0) \\
s(x_1) &= a_{-1}B_{-1}(x_1) + a_0 B_0(x_1) + \cdots + a_{N+1}B_{N+1}(x_1) = f(x_1) \\
&\quad \cdots\cdots\cdots\cdots\cdots\cdots\cdots\cdots\cdots\cdots\cdots\cdots\cdots\cdots\cdots\cdots\cdots\cdots \\
&\quad \cdots\cdots\cdots\cdots\cdots\cdots\cdots\cdots\cdots\cdots\cdots\cdots\cdots\cdots\cdots\cdots\cdots\cdots \\
s(x_N) &= a_{-1}B_{-1}(x_N) + a_0 B_0(x_N) + \cdots + a_{N+1}B_{N+1}(x_N) = f(x_N) \\
s'(x_N) &= a_{-1}B_{-1}(x_N) + a_0 B_0(x_N) + \cdots + a_{N+1}B_{N+1}(x_N) = f'(x_N)
\end{aligned}
$$

$$(3.8)$$

where $a_{-1}, a_0, a_1, \ldots, a_{N+1}$ are unknown coefficients of the cubic spline $s(x)$.

The system of equations (3.8) has a unique solution, since the matrix

$$
M = \left\{
\begin{array}{ccccccccc}
-\dfrac{3}{h} & 0 & \dfrac{3}{h} & 0 & \cdots & 0 & 0 & 0 \\
1 & 4 & 1 & 0 & \cdots & 0 & 0 & 0 \\
0 & 1 & 4 & 1 & \cdots & 0 & 0 & 0 \\
\cdots & \cdots & \cdots & \cdots & \cdots & \cdots & \cdots & \cdots \\
0 & 0 & 0 & 0 & \cdots & 1 & 4 & 1 \\
0 & 0 & 0 & 0 & \cdots & -\dfrac{3}{h} & 0 & \dfrac{3}{h}
\end{array}
\right\}
$$

is non-singular.

The following theorem holds (*cf.* [1, 2, 9]):

Theorem 3.4 *If $s(x)$ is interpolating cubic spline to a function $f(x)$ four times continuously differentiable on the interval $[a, b]$, then the error of interpolation satisfies the following inequality:*

$$
\| f^{(r)} - s^{(r)} \|_\infty \leq \epsilon_r h^{4-r} \| f^{(4)} \|_\infty, \quad r = 0, 1, 2, 3;
$$

where

$$
\epsilon_0 = \frac{5}{384}, \quad \epsilon_1 = \frac{1}{216}(9+\sqrt{3}) \quad \epsilon_2 = \frac{1}{12}(3\sigma+1), \quad \epsilon_3 = \frac{1}{2}(\sigma^2+1),
$$

$$
h = \max_{1 \leq i \leq N}(x_i - x_{i-1}), \quad \| f \|_\infty = \inf_{\mu(\Omega)=0} \sup_{a \leq x \leq b} | f(x) |,
$$

$\mu(\Omega)$ is the measure of the set Ω, and σ is the constant which defines the normal partition of the interval $[a, b]$, and $\sigma = 1$ for the uniform partition of the interval $[a, b]$

Now, let us determine the coefficients $a_{-1}, a_0, a_1, ..., a_{N+1}$ of the cubic spline $s(x)$. Using the table, we can reduce the system of equations (3.8). Namely, from the first equation, we find

$$
a_{-1} = a_1 - \frac{h}{3}f'(x_0),
$$

and from the last equation, we find

$$
a_{N+1} = a_{N-1} + \frac{h}{3}f'(x_N).
$$

Then, the other $N + 1$, equations, we write in the following form:

$$2a_0 + a_1 = \frac{1}{2}[f(x_0) + \frac{h}{3}f'(x_0)]$$

$$a_{i-1} + 4a_i + a_{i+1} = f(x_i), \qquad\qquad i = 1, 2, ..., N - 1,$$

$$a_{N-1} + 2a_N = \frac{1}{2}[f(x_N) - \frac{h}{3}f'(x_N)]$$

In order to solve this system of equations with tri-diagonal matrix

$$M = \left\{ \begin{array}{cccccccc} 2 & 1 & 0 & 0 & \cdots & 0 & 0 & 0 \\ 1 & 4 & 1 & 0 & \cdots & 0 & 0 & 0 \\ 0 & 1 & 4 & 1 & \cdots & 0 & 0 & 0 \\ \cdots & \cdots & \cdots & \cdots & \cdots & \cdots & \cdots & \cdots \\ 0 & 0 & 0 & 0 & \cdots & 1 & 4 & 1 \\ 0 & 0 & 0 & 0 & \cdots & 0 & 0 & 2 \end{array} \right\}_{(N+1)(N+1)}$$

we apply Gauss elimination method for the vector

$$F = \{F_0, F_1, ..., F_N\}$$

in right sides:

$$F_i = \begin{cases} \dfrac{1}{2}(f(x_0) + \dfrac{h}{3}f'(x_0)), & i = 0, \\[2mm] f(x_i), & i = 1, 2, ..., N - 1, \\[2mm] \dfrac{1}{2}(f(x_N) - \dfrac{h}{3}f'(x_N)), & i = N \end{cases}$$

Then, we apply the following algorithm:

$$\alpha_0 = \frac{1}{2}, \qquad\qquad \beta_0 = \frac{1}{2}F_0,$$

$$for \quad i = 1, 2, ..., N - 1,$$

$$\alpha_i = \frac{1}{4 - \alpha_{i-1}}, \qquad\qquad \beta_i = \frac{F_i - \beta_{i-1}}{4 - \alpha_{i-1}},$$

$$\alpha_N = \frac{1}{2 - \alpha_{N-1}}, \qquad\qquad \beta_N = \frac{F_N - \beta_{N-1}}{2 - \alpha_{N-1}},$$

$$a_N = \beta_N,$$

$$for \quad i = N - 1, N - 2, ..., 1, 0,$$

$$a_i = \beta_i - \alpha_i a_{i+1},$$

$$a_{-1} = a_1 - \frac{h}{3}f_0', \qquad\qquad a_{N+1} = a_{N-1} + \frac{h}{3}f_N'.$$

Below, we give the **Mathematica** module based on the above algorithm which produces the table of the cubic spline for a give $f(x)$ in the interval $[a, b]$. How to invoke the module, we shall explain in the example.

Program 3.2 *Mathematica module that finds a cubic spline in the form of a table for given data table.*

```
cubicSpline[f_,a_,b_,n_,tstep_]:=Module[{h,sol,sp3,
                                          onet,cub},
    h=(b-a)/n;
    sol=sp3[h];

    sp3[h_]:=Module[{xi,f1a,f1b,fx,al,be,sa,sb,s},
      xi=Table[a+i*h,{i,0,n}];
      fx=N[Map[f,xi]];

      df[x_]:=D[f[x],x];
      f1a=N[df[x]/.x->a];
```

```
f1b=N[df[x]/.x->b];

fx[[1]]=(fx[[1]]+h*f1a/3)/2;
fx[[n+1]]=(fx[[n+1]]-h*f1b/3)/2;

al[1]=1/2;
al[i_]:=al[i]=If[i<n+1,1/(4-al[i-1]),
         N[1/Sqrt[3]]];

be[1]=fx[[1]]/2;
be[i_]:=be[i]=If[i<n+1,(fx[[i]]
                 -be[i-1])/(4-al[i-1]),
(fx[[n+1]]-be[n])/Sqrt[3]];

s[n+1]=be[n+1];
s[i_]:=s[i]=be[i]-al[i]*s[i+1];

sol=N[Table[s[i],{i,1,n+1}]];
sa=sol[[2]]-h*f1a/3;
PrependTo[sol,sa];
sb=sol[[n+1]]+h*f1b/3;
AppendTo[sol,sb]
];

onet[t_]:=Module[{ },

k=Floor[(t-a)/h+2];
N[ ((xr[[k+1]]-t)^3*sol[[k-1]]+
(h^3+3*h^2*(xr[[k+1]]-t)
           +3*h*(xr[[k+1]]-t)^2
           -3*(xr[[k+1]]-t)^3)*sol[[k]]
           +(h^3+3*h^2*(t-xr[[k]]))
           +3*h*(t-xr[[k]])^2
           -3*(t-xr[[k]])^3)*sol[[k+1]]
           +(t-xr[[k]])^3*sol[[k+2]])/h^3
];
```

```
Print["Cubic spline approximating f(x) "];
Print["Coefficients of the cubic spline :",
                            Take[sol,n+3]];
Print["---------------------------------"];
   xr=Table[a+r*h,{r,-1,n+1}];
cub=Table[{N[t],onet[t],N[f[t]]}, {t,a,b-tstep,
                                 tstep}];
AppendTo[cub,{N[b], N[((xr[[n+2]]-b)^3*sol[[n]]+
       (h^3+3*h^2*(xr[[n+2]]-b)+
       3*h*(xr[[n+2]]-b)^2-3*(xr[[n+2]]-b)^3)*
                            sol[[n+1]]+
       (h^3+3*h^2*(b-xr[[n+1]])+
       3*h*(b-xr[[n+1]])^2-3*(b-xr[[n+1]])^3)*
                            sol[[n+2]]+
    (b-xr[[n+1]])^3*sol[[n+3]])/h^3,N[f[b]]}];
  TableForm[PrependTo[cub,{" x "," cubic ",f[x]}]]
]
```

Example 3.5 *Find a cubic interpolating spline for the following function:*

$$f(x) = e^x, \quad 0 \le x \le 2,$$

spanned on the knots: $x_0 = 0$, $x_1 = 0.5$, $x_2 = 1$, $x_3 = 1.5$, *and* $x_4 = 2$. *Determine an approximate value* $f(1.4)$. *Estimate the error of interpolation for the function* $f(x)$.

Solution. The interpolating cubic spline is

$$s(x) = a_{-1}B_{-1}(x) + a_0 B_0(x) + a_1 B_1(x) + \cdots + a_5 B_5(x),$$

The coefficients a_{-1}, a_0, a_1, a_2, a_3, a_4 and a_5 are determined by the following system of linear equations:

$$-6a_{-1} + 6a_1 = 1,$$

$$a_{i-1} + 4a_i + a_{i+1} = e^{ih}, \quad i = 0.1, 2, 3, 4; \quad h = 0.5,$$

$$-6a_3 + 6a_5 = e^2.$$

Solving this system of equations, we find

$$s(x) = 0.0969067B_{-1}(x) + 0.159880B_0(x) + 0.263573B_1(x)$$

$$+0.434548B_2(x) + 0.7116517B_3(x) + 1.18107B_4(x)$$

$$+1.94803B_5(x),$$

and

$$s(1.4) = 4.05511, \qquad f(1.3) = e^{1.4} = 4.055199967.$$

So that the error of interpolation

$$f(1.4) - s(1.4) = 0.0000899668.$$

satisfies the estimate given in the theorem, *i.e.*,

$$0.0000899668 = |e^{1.4} - s(1.4)| \leq \frac{5}{384}0.5^4\, e^2 = 0.006,$$

Also, we can find the cubic interpolating spline to $f(x) = e^x$ using the `Mathematica` module by executing the following instructions

```
f[x_]:=Exp[x];
cubicSpline[f,0,1,4,0.1];
```

Then, we obtain the following table:

```
cubicSpline[f,0,1,4,0.1]
Cubic spline approximating f(x)
Coefficients of the cubic spline

{0.128454, 0.16494, 0.211787, 0.271938,
 0.349184, 448327, 0.575707}
------------------------------------------------
Out[3]/TableForm=
```

x	cubic	Exp(x)
0	1	1

0.1	1.10316	1.10517
0.2	1.2214	1.2214
0.3	1.34985	1.34986
0.4	1.49181	1.49182
0.5	1.64872	1.64872
0.6	1.82211	1.82212
0.7	2.01376	2.01375
0.8	2.22551	2.22554
0.9	2.45952	2.4596
1.	2.71828	2.71828

One can draw the graph of the cubic spline by the following `Mathematica` instructions

```
data={{0,1},{0.1,1.10517},{0.2,1.2214},
{0.3,1.34986},{0.4,1.49182},{0.5,1.64872},
{0.6,1.82212},{0.7, 2.01375},
{0.8,2.22554},{0.9,2.4596}, {1.0,2.71828}};
Show[Graphics[{Line[data],Spline[data,Cubic]},
                          PlotRange->All]];
```

Program 3.3 *Mathematica module that finds a cubic spline in the form of a list of piecewise cubic polynomials for a given data table.*

```
cubicSpline[f_,a_,b_,n_]:=
   Module[{h,sol,sp3,onet,cub,xr,k,r},
   h=(b-a)/n;
   sol=sp3[h];

   sp3[h_]:=Module[{xi,f1a,f1b,fx,al,be,sa,sb,s},
     xi=Table[a+i*h,{i,0,n}];
     fx=N[Map[f,xi]];

     df[x_]:=D[f[x],x];
     f1a=N[df[x]/.x->a];
     f1b=N[df[x]/.x->b];
```

```
    fx[[1]]=(fx[[1]]+h*f1a/3)/2;
    fx[[n+1]]=(fx[[n+1]]-h*f1b/3)/2;

    al[1]=1/2;
    al[i_]:=al[i]=If[i<n+1,1/(4-al[i-1]),
            N[1/Sqrt[3]]];

    be[1]=fx[[1]]/2;
    be[i_]:=be[i]=If[i<n+1,
                    (fx[[i]]-be[i-1])/(4-al[i-1]),
    (fx[[n+1]]-be[n])/Sqrt[3]];

    s[n+1]=be[n+1];
    s[i_]:=s[i]=be[i]-al[i]*s[i+1];

    sol=N[Table[s[i],{i,1,n+1}]];
    sa=sol[[2]]-h*f1a/3;
    PrependTo[sol,sa];
    sb=sol[[n+1]]+h*f1b/3;
    AppendTo[sol,sb]
  ];
  onet[t_,k_]:=Module[{ },
    ((xr[[k+1]]-t)^3*sol[[k-1]]+
    (h^3+3*h^2*(xr[[k+1]]-t)+
    3*h*(xr[[k+1]]-t)^2-3*(xr[[k+1]]-t)^3)*sol[[k]]+
    (h^3+3*h^2*(t-xr[[k]])+
    3*h*(t-xr[[k]])^2-3*(t-xr[[k]])^3)*sol[[k+1]]+
    (t-xr[[k]])^3*sol[[k+2]])/h^3
  ];
  xr=Table[a+r*h,{r,-1,n+1}];
  cub=Table[Expand[onet[t,k]], {k,2,n+1}]
  ]
```

Solving the example 4.4, we find the cubic interpolating spline to the function $f(x) = e^x, x \in [0, 2]$, by executing the following instructions

```
f[x_]:=Exp[ x];
cubicSpline[f,0,2,4]
```

Then, we obtain the following list of piecewise cubic splines
determined on the subintervals $[0, 0.5]$, $[0.5, 1]$, $[1, 1.5]$ and
$[1.5, 2]$.

$$\{1. + 1.\,t + 0.48864\,t^2 + 0.21249\,t^3,$$
$$0.982848 + 1.10291\,t + 0.282818\,t^2 + 0.349704\,t^3,$$
$$0.759801 + 1.77205\,t - 0.386323\,t^2 + 0.572752\,t^3,$$
$$-0.541948 + 4.37555\,t - 2.12199\,t^2 + 0.958444\,t^3\}$$

3.4   Lagrange Interpolating Splines

In the previous section, we have introduced splines of the
space $S_m(\Delta, m - 1)$ giving explicit form of the basis. These
splines are defined on a set of spline knots $\{x_0, x_1, ..., x_n\}$.
The Lagrange's interpolating splines are determined by Lagrange's conditions of interpolation given at the Lagrange's
interpolating points $\{t_0, t_1, ..., t_p\}$. In general, these two sets
of points do not coincide. However, in the case of piecewise
linear splines both sets of points are the same. In order to
determine higher order polynomial splines, by Lagrange's
interpolating conditions, one has to consider the number of
conditions equal to the number of coefficients in a piecewise
polynomial. For example, to determine quadratic piecewise
splines $a_i x^2 + b_x + c_i$, $i = 0, 1, ..., n - 1$, by Lagrange's conditions of interpolation, we have to find $3n$ coefficients for
which $3n$ conditions are needed. In the following example,
we shall give the Lagrange's interpolating quadratic splines
spanned at the spline knots $\{x_0, x_1, ..., x_n\}$ and with two
different sets of interpolating points.

Example 3.6 *Find the piecewise quadratic polynomial spline* $P_2^{(i)}(x)$ *spanned on knots* $x_0, x_1, ..., x_n$, *which satisfies La-grange's conditions of interpolation*

$$P_2^{(i)}(x_i) = f(x_i), \quad i = 0, 1, ..., n,$$

for a given function $f(x)$ *in the interval* $[x_0, x_n]$.

Such a spline must be continuous and continuously differentiable in the interval $[x_0, x_n]$. First, let us consider the interpolating points the same as spline knots, that is, $t_i = x_i$, $i = 0, 1, ..., n$. By the Newton's formula, we find the piecewise quadratic polynomial

$$P_2^{(i)}(x) = f_i + \frac{f_{i+1} - f_i}{x_{i+1} - x_i}(x - x_i) + a_i(x - x_i)(x - x_{i+1}),$$

which satisfies the Lagrange's conditions for any values of parameters a_i, and for given $f_i = f(x_i)$, $i = 0, 1, ..., n$. The parameters a_i, $i = 0, 1, ..., n - 1$, are determined by the following continuity conditions

$$\lim_{x \to x_i^-} \frac{dP_2^{(i-1)}(x)}{dx} = \lim_{x \to x_i^+} \frac{dP_2^{(i)}(x)}{dx}, \quad i = 1, 2, ..., n - 1.$$
$$(3.9)$$

Hence, we obtain the equations

$$a_i(x_{i+1} - x_i) + a_{i-1}(x_i - x_{i-1}) = \frac{f_{i+1} - f_i}{x_{i+1} - x_i} - \frac{f_i - f_{i-1}}{x_i - x_{i-1}}.$$

In the case when $x_i = x_0 + ih$, $i = 0, 1, ..., n$, these equations lead to the following recursive formula

$$a_i = -a_{i-1} + \frac{1}{h^2}[f_{i+1} - 2f_i + f_{i-1}], \quad i = 1, 2, ..., n - 1.$$

at which the initial value of the parameter a_0 is free, and can be arbitrarily chosen.

The following **Mathematica** module spanned on knots

$$x_0, x_1, ..., x_n$$

tabulates the quadratic spline that approximates a given function $f(x)$

Program 3.4 *Mathematica module that tabulates the quadratic spline approximating a given function.*

```
quadraticSpline1[f_,a_,b_,n_,tstep_]:=
                Module[{h,xi,fx,k,t},
   h=(b-a)/n;
   xi=Table[a+i*h,{i,0,n+1}];
   fx=Map[f,xi];
   al[1]=1;
   al[i_]:=al[i]=-al[i-1]+(fx[[i-1]]-2 fx[[i]]
                      +fx[[i+1]])/h^2;
 onet[t_]:=Module[{ },
  k=Floor[(t-a)/h+1];
  N[f[a+(k-1)*h]+(f[a+(k+1)*h]
                -f[a+k*h])*(t-xi[[k]])/h
                +al[k]*(f[a+(k-1)*h]-2 f[a+k*h]
      +f[a+(k+1)*h])*(t-xi[[k]])*(t-xi[[k+1]])]
     ];

   Print["Quadratic1 spline approximating f(x)"];
   Print["---------------------------------"];
   quad=Table[{N[t],onet[t],N[f[t]]},{t,a,b-tstep,
                                     tstep}];
   AppendTo[quad,{N[b],onet[b],N[f[b]]}];
   TableForm[PrependTo[quad,{"x"," quadratic1 ",
                                     f[x]}]]
   ]
```

How to use the Mathematica module, we explain by the following example:

Example 3.7 *Let*

$$f(x) = e^x, \qquad 0 \le x \le 1,$$

Tabulate the quadratic spline with step tstep = 0.1 which is spanned on knots $x_i = i\,h$, $i = 0, 1, 2, 3, 4, 5$, and approximates $f(x)$

When `Mathematica` is active, we enter the function

`f[x_]:=Exp[x];`

and call

`quadraticSpline1[f,0,1,5,0.1];`

where $a = 0$, $b = 1$, number of intervals $n = 5$, and $tstep = 0.1$.

Then, obtain the following table:

```
quadraticSpline1[f,0,1,5,0.1]
Out[3]/TableForm=
quadratic spline approximating f(x)
------------------------------------
```

x	quadratic1	e^x
0	1.	1.
0.1	1.134721	1.105171
0.2	1.221403	1.221403
0.3	1.386415	1.349859
0.4	1.491825	1.491825
0.5	1.692601	1.648721
0.6	1.822119	1.822119
0.7	2.067992	2.013753
0.8	2.225541	2.225541
0.9	2.524663	2.459601
1.	2.718282	2.718282

Now, let us consider $n + 2$ interpolating points

$$
t_i = \begin{cases}
x_0, & if \quad i = 0, \\
\dfrac{1}{2}(x_{i-1} + x_i), & if \quad i = 1, 2, ..., n, \\
x_n & if \quad i = n + 1.
\end{cases}
$$

at which $f_i = f(t_i)$, $i = 0, 1, ..., n + 1$.

For the equidistance points $x_i = a + ih$, $i = 0, 1, ..., n$, $nh = b - a$, the Newton's quadratic interpolating polynomial through the points t_i, t_{i+1}, t_{i+2} is

$$P_2^{(i)}(x) = f_i + \frac{1}{h}(f_{i+1} - f_i)(x - t_i)$$

$$+ \frac{1}{2h^2}[f_i - 2f_{i+1} + f_{i+2}](x - t_i)(x - t_{i+1}),$$

$$x_i \leq x < x_{i+1}, \quad i = 1, 2, ..., n - 2,$$

This polynomial satisfies the Lagrange's interpolating conditions

$$P_2^{(i)}(t_i) = f_i,$$

$$P_2^{(i)}(t_{i+1}) = f_{i+1},$$

$$P_2^{(i)}(t_{i+2}) = f_{i+2}, \quad i = 1, 2, ..., n - 2.$$

and the continuity conditions (3.9).

Additionally, we need to find two quadratic splines that correspond to the points t_0, t_1, and t_n, t_{n+1} which satisfy the continuity conditions at the points x_1 and x_{n-1}. One can check that the following quadratic polynomials hold these conditions (**see Fig. 3.4**)

$$P_2^{(0)}(x) = f_0 + \frac{2}{h}(f_1 - f_0)(x - t_0)$$

$$+ \frac{2}{3h^2}(2f_0 - 3f_1 + f_2)(x - t_0)(x - t_1),$$

$$x_0 \leq x < x_1,$$

and

$$P_2^{(n-1)}(x) = f_n + \frac{2}{h}(f_{n+1} - f_n)(x - t_n) +$$

$$+ \frac{2}{3h^2}(f_{n-1} - 3f_n + 2f_{n+1})(x - t_n)(x - t_{n+1}),$$

$$x_{n-1} \leq x \leq x_n.$$

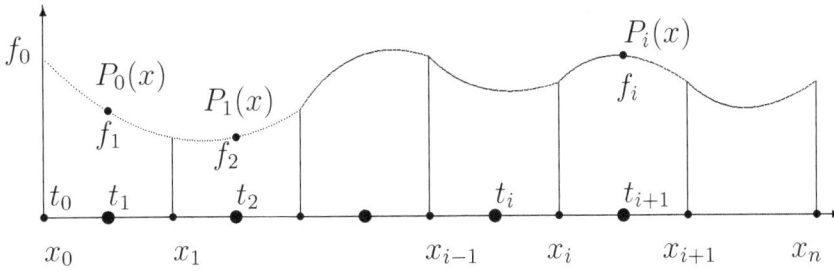

Fig. 3.4. Quadratic Spline

The error of interpolation is given by the formula

$$f(x) - P_2^{(i)}(x) = \frac{f'''(\xi_i)}{3!}(x - t_i)(x - t_{i+1})(x - t_{i+2}).$$

Hence, we have the following error estimate

$$|f(x) - P_2^{(i)}(x)| \leq \frac{M^{(3)}\sqrt{3}}{27}h^3, \ x_i \leq x \leq x_{i+1}, \ i = 0, 1, ..., n-1,$$

where $M^{(3)} = \sup\limits_{x}|f^{('')}(x)|$.

The following module in **Mathematica** tabulates the quadratic spline spanned on the Lagrange's interpolating knots

$$\{t_0, t_1, ..., t_n\}$$

that approximates a function $f(x)$.

Program 3.5 *Mathematica module that tabulates the Lagrange's quadratic spline approximating a given function.*

```
quadraticSpline2[f_,a_,b_,n_,tstep_]:=Module[{h},
    h=N[(b-a)/n];
    xi=Table[N[a+i*h],{i,0,n}];
    xt=Table[N[(xi[[i]]+xi[[i+1]])/2],{i,1,n}];
```

```
PrependTo[xt,a];
AppendTo[xt,b];

p0[t_]:=Module[{ },
        f[a]+2*(f[xt[[2]]]-f[a])*(t-xt[[1]])/h+
        2*(2*f[a]-3*f[xt[[2]]]+
        f[xt[[3]]])*(t-a)*(t-xt[[2]])/(3*h^2)
        ];
pi[t_]:=Module[{ },
          r=Floor[N[(t-a)/h]];
          f[xt[[r+1]]]+(f[xt[[r+2]]]-
          f[xt[[r+1]]])*(t-xt[[r+1]])/h+
          (f[xt[[r+1]]]-2*f[xt[[r+2]]]+
           f[xt[[r+3]]])*(t-xt[[r+1]])*
           (t-xt[[r+2]])/(2*h^2)
          ];
pn[t_]:=Module[{ },
    f[xt[[n+1]]]+2*(f[xt[[n+2]]]-
    f[xt[[n+1]]])*(t-xt[[n+1]])/h+
    2*(2*f[xt[[n]]]-3*f[xt[[n+1]]]+f[xt[[n+2]]])*
    (t-xt[[n+1]])*(t-xt[[n+2]])/(3*h^2)
         ];
onet[t_]:=Module[{ },
        Which[t<=xt[[2]],N[p0[t]],
        t>=xt[[n+1]],N[pn[t]],True,N[pi[t]]]
        ];

Print["Quadratic2 spline approximating f(x) "];
Print[" --------------------------------"];

quad=Table[{N[t],onet[t],N[f[t]]},{t,a,b,tstep}];

TableForm[PrependTo[quad,{" x ","quadratic2 ",f[x]}]]
]
```

Coming back to the example 4.6, we invoke the module by
the instructions

```
f[x_]:=Exp[x];
quadraticSpline2[f,0,1,5,0.1]
```

Then, we obtain the following table

```
quadraticSpline2[f,0,1,5,0.1]
Quadratic spline approximating f(x)
--------------------------------
```

```
Out[2]/TableForm=
```

x	quadratic2	e^x
0	1.	1.
0.1	1.10517	1.10517
0.2	1.22074	1.2214
0.3	1.34986	1.34986
0.4	1.49102	1.49182
0.5	1.64872	1.64872
0.6	1.82113	1.82212
0.7	2.01375	2.01375
0.8	2.22658	2.22554
0.9	2.4596	2.4596
1.	2.71828	2.71828

3.5 Polynomial Splines of Two Variables on Rectangular Networks

Let us first consider polynomial splines associated with a rectangular network $\Delta = (\Delta_x, \Delta_y)$ which are defined an the rectangle

$$R = (x, y) \; : \; a \leq x \leq b, \;\; c \leq y \leq d,$$

where

$$\Delta x \; : \;\; a = x_0 < x_1 < \cdots < x_{N_1} = b,$$
$$\Delta y \; : \;\; c = y_0 < y_1 < \cdots < y_{N_2} = d,$$

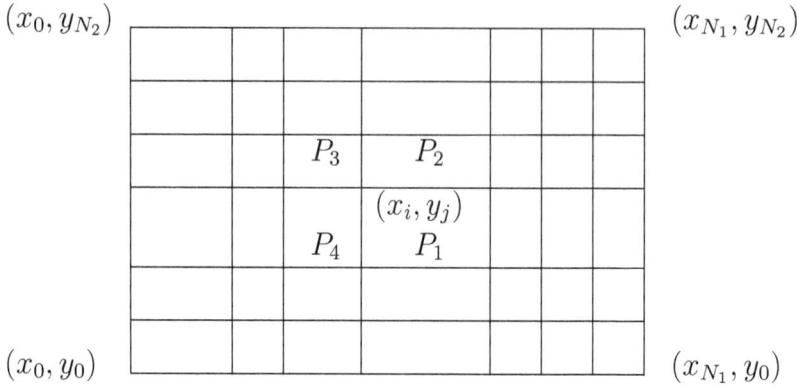

Fig. 3.5. Δ partition

Definition 3.2 *A function $s(x, y)$ is said to be a polynomial spline defined on a rectangular network if the following conditions are satisfied:*

    (i) $s; \in S_{m_1}(\Delta_x, k_1)$ *for any* $y \in [c, d]$,

    (ii) $s \in S_{m_2}(\Delta_y, k_2)$ *for any* $x \in [a, b]$,

where $S_{m_1}(\Delta_x, k)$ and $S_{m_2}(\Delta_y, k_2)$ are polynomial spaces of one variable (def. 4.1) In symbols, we write $s \in S_{m_1, m_2}(\Delta, k_1, k_2)$.

Space $S_{11}(\Delta, 0, 0)$ of bilinear splines. As the base of the space $S_{11}(\Delta, 0, 0)$ we can consider the following products:

$$\psi_{ij}(x, y) = \psi_i(x)\psi_j(y), \quad i = 0, 1, \ldots, N_1, \quad j = 0, 1, \ldots, N_2,$$

where $(x, y) \in R$,

$$\psi_0(x), \ \psi_1(x), \ \ldots, \psi_{N1}(x)$$

is the basis of the space $S_1(\Delta_x, 0)$,
and

$$\psi_0(y), \ \psi_1(y), \ \ldots, \psi_{N_2}(y)$$

is the basis of the space $S_1(\Delta_y, 0)$.

The explicit form of the spline $\psi_{ij}(x, y)$ is given below (*cf.* [21]):

$$
\psi_{ij}(x, y) = \begin{cases}
\dfrac{x_{i+1} - x}{x_{i+1} - x_i} \dfrac{y - y_{j-1}}{y_j - y_{j-1}} & \text{in } P_1, \\[2ex]
\dfrac{x_{i+1} - x}{x_{i+1} - x_i} \dfrac{y_{j+1} - y}{y_{j+1} - y_j} & \text{in } P_2, \\[2ex]
\dfrac{x - x_{i-1}}{x_i - x_{i-1}} \dfrac{y_{j+1} - y}{y_{j+1} - y_j} & \text{in } P_3, \\[2ex]
\dfrac{x - x_{i-1}}{x_i - x_{i-1}} \dfrac{y - y_{j-1}}{y_j - y_{j-1}} & \text{in } P_4, \\[2ex]
0 & \text{in } R - (P_1 \cup P_2 \cup P_3 \cup P_4).
\end{cases}
$$

for $i = 0, 1, \ldots, N_1, \quad j = 0, 1, \ldots, N_2$.

Thus, we can write any piecewise linear spline given on mesh grids (**see fig 3.5**) network Δ in the form of the following linear combination:

$$
s(x, y) = \sum_{i=0}^{N_1} \sum_{j=0}^{N_2} a_{ij} \psi_{ij}(x, y).
$$

Interpolation in the space $S_{11}(\Delta, 0, 0)$. Let $f(x, y)$ be a given function twice continuously differentiable in the closed rectangle R. The interpolating spline $s \in S_{11}(x, y)$ which satisfies the conditions:

$$
s(x_i, y_j) = f(x_i, y_j); \quad i = 0, 1, \ldots, N_1; \quad j = 0, 1, \ldots, N_2,
$$

has the following form:

$$
s(x, y) = \sum_{i=0}^{N_1} \sum_{j=0}^{N_2} f(x_i, y_j) \psi_{ij}(x, y).
$$

By the theorem on interpolation (*cf.* [9, 21]), the error $s(x, y) - f(x, y)$ satisfies the following inequality:

$$
\| s^{(r)} - f^{(r)} \| \leq \alpha_r h^{2-r}, \quad r = 0, 1,
$$

where $h = \max\{h_x, h_y\}$, $h_x = \max\{x_{i+1} - x_i\}$, $h_y = \max\{y_{j+1} - y_j\}$; α_r, $r = 0, 1$,   are constants independent of h.

Space $S_{33}(\Delta, 2, 2)$ of bicubic splines. The space $S_{33}(\Delta, 2, 2)$ is determined by the following linearly independent set of cubic splines:

$$B_{ij}(x, y) = B_i(x) B_j(y),$$

$$i = -1, 0, 1, \ldots, N_1+1, \quad j = -1, 0, 1, \ldots, N_2+1, \quad (x, y) \in R,$$

where $B_i(x)$, $i = -1, 0, 1, \ldots, N_1 + 1$, and $B_j(y)$, for $j = -1, 0, 1, \ldots, N_2+1$ are the corresponding basis of the spaces $S_3(\Delta_x, 2)$ and $S_3(\Delta_y, 2)$. Therefore, any bicubic spline given on the rectangular network Δ has the following form:

$$s(x, y) = \sum_{i=-1}^{N_1+1} \sum_{j=-1}^{N_2+1} a_{ij} B_{ij}(x, y).$$

Interpolation in the space $S_{33}(\Delta, 2, 2)$. Let $f(x, y)$ be a function four times continuously differentiable in the closed rectangle R. Then, the interpolating bicubic spline $s(x, y)$ to the function $f(x, y)$ is uniquely determined by the following conditions:

$$s(x_i, y_j) = f(x_i, y_j),$$

$$i = -1, 0, 1, \ldots, N_1 + 1; \quad j = -1, 0, 1, \ldots, N_2 + 1,$$

$$\frac{\partial s}{\partial x}(x_i, y_j) = \frac{\partial f}{\partial x}(x_i, y_j), \quad i = 0, N_1, \quad \text{when} \quad j = 0, 1, \ldots, N_2,$$

$$\frac{\partial^2 s}{\partial x \partial y}(x_i, y_j) = \frac{\partial^2 f}{\partial x \partial y}(x_i, y_j), \quad i = 0, N - 1, \quad \text{and} \quad j = 0, N_2.$$

By the theorem on interpolation (*cf.* [21]), the error $s(x, y) - f(x, y)$ satisfies the following inequalities:

$$\| s - f \|_\infty \leq \beta_) h^4,$$

$$\| \frac{\partial (s - f)}{\partial x} \|_\infty \leq \beta_0 h^3, \quad \| \frac{\partial (s - f)}{\partial y} \| \leq \beta_1 h^3,$$

$$\| \frac{\partial^2 (s - f)}{\partial x^2} \|_\infty \leq \beta_2 h^2, \quad \| \frac{\partial^2 (s - f)}{\partial y^2} \| \leq \beta_3 h^2,$$

$$\| \frac{\partial^2 (s - f)}{\partial x \partial y} \|_\infty \leq \beta_4 h^2,$$

where $\beta_i, \quad i = 0, 1, 2, 3, 4$ are constants independent of h.

3.6   Space Π_1 of Piecewise Linear Polynomial Splines on Triangular Networks

The piecewise linear splines on a triangular network have a simple structure and they are applicable to domains of arbitrary shape (*cf.* [29]). Let us consider the following triangularization of a bounded domain Ω:

$$\Delta = \{T_0, T_1, \ldots, T_n\},$$

where

(*i*) $\overline{\Omega} = \overline{T_0} \cup \overline{T_1} \cup \cdots \cup \overline{T_n}$
(*ii*) $\overline{T_i} \cap \overline{T_j} = \emptyset$ *or they have a common side or vertix if* $i \neq j$,
(*iii*) *there exists a positive constant* ν_0 *independent of n*
such that
$\nu = \min_{\alpha_T, \beta_T, \gamma_T} \geq \nu_0,$ *for all* $T \in \Delta,$

here α_T, β_T, and γ_T are the angles of the triangle T.

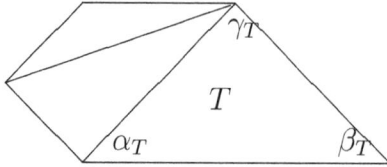

Fig. 3.6. Triangular grids

Let
$$p_k(x, y) = a_k x + b_k y + c_k, \quad (x, y) \in T_k,$$
be a piece of a linear function and let $\Pi_1(\Delta)$ be the set of all continuous piecewise linear functions of the form:

$$s(x, y) = p_k(x, y), \quad \text{for} \quad (x, y) \in T_k, \quad k = 0, 1, \dots, n.$$

By the theorem on interpolation (*cf.* [29]), there exists a unique interpolating piecewise linear spline $s \in \Pi_1(\Delta)$ to a function $f(x, y)$ which satisfies the following conditions:

$$s(x_{kl}, y_{kl}) = f(x_{kl}, y_{kl}), \quad \text{for} \quad l = 0, 1, 2,$$

where (x_{k0}, y_{k0}), (x_{k1}, y_{k1}) and (x_{k2}, y_{k2}) are vertices of the triangle T_k for $k = 0, 1, \dots, n$.
This spline approximates $f(x, y)$ with the error

$$s(x, y) - f(x, y)$$

which satisfies the inequality (*cf.* [29])

$$\| s - f \|_\infty \le \beta h^2,$$

where $\beta > 0$ is a positive constant independent of h,

$$h = \max_{T \in \Delta} \ \max\{l_T^0, l_T^1, l_T^2\},$$

and l_T^0, l_T^1, l_T^2 are sides of the triangle T, (see **Fig. 3.6**).
Space $\Pi_3(\Delta)$ of Cubic Splines on Triangular Networks. Let us consider a piece of a cubic polynomial in

the following form:

$$P_T(x, y) = \alpha_1 + \alpha_2 x + \alpha_3 y + \alpha_4 x^2 + \alpha_5 xy + \alpha_6 y^2 + \alpha_7 x^3$$
$$+ \alpha_8 x^2 y + \alpha_9 xy^2 + \alpha_{10} y^3, \quad (x, y) \in T.$$

We shall use $\Pi_3(\Delta)$ to denote the set of all piecewise continuous cubic splines such that

$$s(x, y) = p_{T_k}(x, y), \quad \text{for } (x, y) \in T_k, \quad k = 0, 1, \ldots, n.$$

In order to find an interpolating cubic spline in the space $\Pi_3(\Delta)$, we shall consider the triangle T with the center at the point Q_0, and its vertices at points Q_1, Q_2 and Q_3, (**see Fig. 3.7**)

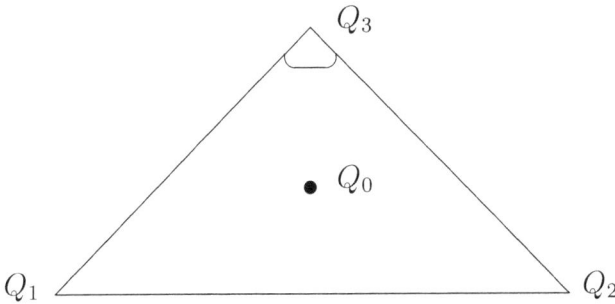

Fig. 3.7. Triangle T

The interpolating piecewise linear cubic spline $p_T(x, y)$ is uniquely determined in the space $\Pi_3(\Delta)$ by the following conditions (*cf.* [29]):

$$P_T(Q_k) = f(Q_k), \quad k = 0, 1, 2, 3,$$

$$\frac{\partial P_T(Q_k)}{\partial x} = \frac{\partial f(Q_k)}{\partial x}, \quad k = 1, 2, 3,$$

$$\frac{\partial P_T(Q_k)}{\partial y} = \frac{\partial f(Q_k)}{\partial y}, \quad k = 1, 2, 3,$$

for every $T \in \Delta$.
The following theorem holds:

Theorem 3.5 *(cf. [29]) If f is a function four times continuously differentiable in the domain Ω, and $p_T \in \Pi_3(\Delta)$ is the linear piecewise interpolating cubic spline to f, then the error of interpolation $s - f$ satisfies the inequalities*

$$\| s - f \|_\infty \leq \frac{3M^{(4)}}{\sin \alpha} h^4,$$

$$\| \frac{\partial(s - f)}{\partial x} \|_\infty \leq \frac{5M^{(4)}}{\sin \alpha} h^3,$$

$$\| \frac{\partial(s - f)}{\partial y} \|_\infty \leq \frac{5M^{(4)}}{\sin \alpha} h^3,$$

where

$$\alpha = \min_{T \in \Delta}, \ \min\{\alpha_T, \beta_T, \gamma_T\} > v_0 > 0, \quad h = \max_{T \in \Delta} \max\{l_T^1, l_T^2, l_T^3\}.$$

3.7 Exercises

Question 3.1 *Approximate the function*

$$f(x) = ln(1 + x), \quad 0 \leq x \leq 2,$$

by a piecewise linear spline with accuracy $\epsilon = 0.01$. Find an approximate value $f(1.3)$.

Question 3.2 *Approximate the function*

$$f(x) = \sqrt{1 + x}, \quad 0 \leq x \leq 2,$$

and its first, second and third derivatives by a cubic spline with accuracy $\epsilon = 0.01$.

Question 3.3 *Give a basis of $S_2(\Delta, 1)$ space. Use this basis to determine a quadratic spline that approximates a function $f(x)$.*

Question 3.4 *Write a module in Mathematica that tabulates a quadratic spline $s \in S_2(\Delta, 1)$ approximating the function $f(x)$.*

Question 3.5 *Give a basis of the space $S_5(\Delta, 4)$ of quintic splines. Use this basis to determine a quintic spline approximating a function $f(x)$.*

Question 3.6 *Write a module in Mathematica that tabulates a quintic spline $s \in S_5(\Delta, 4)$ approximating the function $f(x)$.*

Question 3.7 *Let $s \in S_3(\Delta, 2)$ be a cubic spline. Following the proof of theorem 4.3, show that s minimizes the functional*

$$F(g) = \int_a^b [g''(x)]^2 dx, \quad g \in C^2[a, b],$$

under the interpolation conditions:

$$g(x_i) = f(x_i), \quad i = 0, 1, ..., N, \quad g'(a) = s'(b) = 0.$$

Send Orders for Reprints to reprints@benthamscience.net
Lecture Notes in Numerical Analysis with Mathematica, 2014, 103-132   **103**

CHAPTER 4

Uniform Approximation

Abstract

In the subject of uniform approximations the fundamental theorems with their proofs are presented. Like Taylor and Weierstrass Theorems, Bernstein's and Chebyshev's polynomials, Chebyshev's series and the best uniform approximation of a function are considered . The theorems are clarified by examples. The Mathematica modules are designed for uniform approximation of one variable functions. The chapter ends with a set of questions.

Keywords: Uniform approximation, Bernstein's polynomials, Chebyshe's polynomials.

4.1 Taylor Polynomials and Taylor Theorem

Let f be a function n-times continuously differentiable in the interval $[a, b]$. Then, the polynomial

$$
\begin{aligned}
TL_n(x) \;=\; & f(x_0) + \frac{f'(x_0)}{1!}(x - x_0) + \frac{f''(x_0)}{2!}(x - x_0)^2 \\
& + \frac{f'''(x_0)}{3!}(x - x_0)^3 + \cdots + \frac{f^{(n)}(x_0)}{n!}(x - x_0)^n,
\end{aligned}
$$

is called *Taylor polynomial of degree* n *of the function* f about the point x_0.[1]
and the numbers

$$\frac{f^{(k)}(x_0)}{n!}, \quad k = 0, 1, \ldots, n,$$

are called *Taylor coefficients* of f.
The relationship between a function and its Taylor polynomial is given in the following Taylor's theorem:

Theorem 4.1 *If* f *is a function* $(n+1)$ *times continuously differentiable in the closed interval* $[a, b]$, *then there exists a point* $\xi_x \in (a, b)$ *such that*

$$
\begin{aligned}
f(x &= f(x_0) + \frac{f'(x_0)}{1!}(x - x_0) + \frac{f''(x_0)}{2!}(x - x_0)^2 \\
&+ \frac{f'''(x_0)}{3!}(x - x_0)^3 + \cdots + + \frac{f^{(n)}}{n!}(x - x_0)^n + R_{n+1}(\xi_x),
\end{aligned}
$$

for all $x_0, x \in [a, b]$, *where the remainder* $R_{(n+1)}(\xi_x)$ *can be written in the following forms:*
(a) *The Lagrange's form:*

$$R_{n+1}(\xi_x) = \frac{f^{(n+1)}(\xi_x)}{(n+1)!}(x - x_0)^{(n+1)}.$$

where ξ_x *is between* x *and* x_0.
(b) *The Cauchy's form:*

$$R_{n+1}(\xi_x) = \frac{f^{(n+1)}(x_0)}{n!}(x - x_0)(x - \xi_x)^n.$$

(c) *The Integral form:*

$$R_{n+1}(\xi_x) = \int_{x_0}^{x} \frac{(x - t)^n}{n!} f^{(n+1)}(t) dt.$$

---

[1]Taylor polynomial about $x_0 = 0$ of f is referred as Maclaurin's polynomial of f.

Proof. We shall prove Taylor theorem with the remainder $R_{n+1}(\xi_x)$ given in the Lagrange's form (**a**). Let us consider the following auxiliary function:

$$g(t) = f(x) - f(t) - \frac{x-t}{1!}f'(t) - \frac{(x-t)^2}{2!}f''(t) \cdots$$
$$- \frac{(x-t)^n}{n!}f^{(n)}(t),$$

(4.1)

for $x, t \in [a, b]$.

Obviously, the derivative $g'(t)$ exists and

$$g'(t) = -f'(t) + f'(t) - \frac{x-t}{1!}f''(t) + \frac{x-t}{1!}f''(t) \cdots$$
$$- \frac{(x-t)^n}{n!}f^{(n+1)}(t) = -\frac{(x-t)^n}{n!}f^{(n+1)}(t).$$

Now, let us consider another auxiliary function

$$G(t) = g(t) - \frac{g(x_0)}{(x-x_0)^{k+1}}(x-t)^{k+1},$$

where t is between x_0 and x.

This function satisfies Rolle's theorem for any integer k, since we have

$$G(x_0) = G(x) = 0,$$

and $G'(t)$ exists in the open interval (a, b). By the Rolle's theorem, there exists a point ξ_x such that

$$G'(\xi_x) = 0.$$

On the other hand

$$G'(t) = g'(t) + \frac{g(x_0)}{(x-x_0)^{k+1}}(k+1)(x-t)^k.$$

Thus, for $k = n$ and $t = \xi_x$, we have

$$-\frac{(x-\xi_x)^n}{n!}f^{(n+1)}(\xi_x) + g(x_0)(n+1)(x-x_0)^{n+1}(x-\xi_x)^n = 0.$$

Hence

$$g(x_0) = \frac{f^{(n+1)}(\xi)}{(n+1)!}(x - x_0)^{n+1}.$$

and by (4.1), we obtain the Taylor formula

$$f(x) = f(x_0) + \frac{f'(x_0)}{1!}(x - x_0) + \frac{f''(x_0)}{2!}(x - x_0)^2 + \cdots$$
$$+ \frac{f^{(n)}}{n!}(x - x_0)^n + R_{n+1}(\xi_x)(x - x_0)^{n+1},$$

where the remainder

$$R_{n+1}(\xi_x) = \frac{f^{(n+1)}(\xi_x)}{(n+1)!}(x - x_0)^{n+1}.$$

Example 4.1 *Find Taylor polynomial and determine the remainder in the Lagrange's form for the function $f(x) = e^x$, $-\infty < x < \infty$, when $x_0 = 0$.*

Solution. In order to determine the Taylor polynomial

$$TL_n(x) = f(x_0) + \frac{f'(x_0)}{1!}(x - x_0) + \frac{f''(x_0)}{2!}(x - x_0)^2 + \cdots$$
$$+ \frac{f^{(n)}}{n!}(x - x_0)^n,$$

we shall find the Taylor coefficients of e^x. Clearly

$$f^{(n)}(x) = e^x \quad \text{for all} \quad n = 0, 1, 2, \ldots.$$

Thus, Taylor polynomial

$$TL_n(x) = 1 + \frac{x}{1!} + \frac{x^2}{2!} + \frac{x^3}{3!} + \cdots + \frac{x^n}{n!},$$

where the Lagrange's remainder

$$R_{n+1}(\xi_x) = \frac{f^{(n+1)}(\xi_x)}{(n+1)!}x^{n+1}.$$

One can obtain the Taylor's polynomial by executing the following instruction in `Mathematica`

$$\texttt{Series[Exp[x],\{x,0,n\}]}$$

for given $n = 0, 1, , , ,$

Example 4.2 *Find Taylor polynomial for the function $f(x) = ln(1+x), \quad 0 \leq x \leq 1$, about $x_0 = 0$. How many terms of the Taylor polynomial are required to approximate the function $ln(1+x), \quad 0 \leq x \leq 1$, by its Taylor polynomial with accuracy $\epsilon = 0.0001$.*

Solution. In order to obtain Taylor polynomial of $f(x) = ln(1+x)$, we calculate

$$f(x) = ln(1+x), \qquad f(0) = 0,$$

$$f'(x) = \frac{1}{1+x}, \qquad f'(0) = 1,$$

$$f''(x) = -\frac{1!}{(1+x)^2}, \qquad f''(0) = -1,$$

$$f'''(x) = \frac{2!}{(1+x)^3}, \qquad f'''(0) = 2,$$

$$f^{(4)}(x) = -\frac{3!}{(1+x)^4}, \qquad f^4(0) = 4!.$$

In general

$$f^{(n)}(x) = (-1)^{n-1} \frac{(n-1)!}{(1+x)^n} \quad \text{and} \quad f^n(0) = (-1)^{n-1}(n-1)!,$$

for $n = 0, 1, 2, \ldots$.
Hence, Taylor polynomial of $ln(1+x)$ at $x_0 = 0$ is

$$TL_n(x) = x - \frac{x^2}{2} + \frac{x^3}{3} - \frac{x^4}{4} + \cdots + (-1)^{n-1}\frac{x^n}{n},$$

where the Lagrange's remainder

$$R_{n+1}(\xi_x) = (-1)^n \frac{x^{n+1}}{(n+1)(1+\xi_x)^{n+1}}.$$

The error of approximation

$$ln(1 + x) - TL_n(x) = (-1)^n \frac{x^{n+1}}{(n+1)(1 + \xi_x)^{n+1}}$$

satisfies the inequality

$$| \ ln(1 + x) - TL_n(x) \ | \leq \frac{1}{n+1}, \qquad 0 \leq x \leq 1.$$

The required number of terms of Taylor polynomial $TL_n(x)$ to get the accuracy $\epsilon = 0.0001$ is determined by the following inequality:

$$\frac{1}{n+1} \leq 0.0001 \quad \text{or} \quad n \geq 9999.$$

We note that the Taylor's series of the function $ln(1 + x)$ is slowly convergent. For example, to compute $\ln 2$ with the accuracy ϵ, we need to add about $[\frac{1}{\epsilon}]$ terms. We can compute this sum by the instructions

```
N[Sum[(-1)^(n+1)/n,{n,1,9999}]]
```

Then, we obtain $\ln 2 \approx 0.693197$, while direct computation from `Mathematica` gives `Log[2]=0.693147`.

Example 4.3 *Consider the following functions:*

1. $f(x) = \sin x, \quad 0 \leq x \leq \frac{\pi}{2}$,

2. $f(x) = \cos x, \quad 0 \leq x \leq \frac{\pi}{2}$.

    (a) Find Taylor polynomial for the above functions at $x_0 = 0$.

    (b) For what value of n will Taylor polynomial approximate the above functions correctly upto three decimal places in the interval $[0, \frac{\pi}{2}]$.

Solution (a). In order to find Taylor polynomial

$$TL_n(x) = f(x_0) + \frac{f'(x_0)}{1!}(x - x_0) + \frac{f''(x_0)}{2!}(x - x_0)^2 + \cdots$$

$$+ \frac{f^{(n)}(x_0)}{n!}(x - x_))^n,$$

we shall determine Taylor coefficients

$$\frac{f^{(k)}(0)}{k!}, \quad k = 0.1, \ldots, n.$$

We have

$$f(x) = sin\ x, \qquad f(0) = 0,$$

$$f'(x) = cos\ x, \qquad f'(0) = 1,$$

$$f''(x) = -sin\ x, \quad f''(0) = 0,$$

$$f'''(x) = -cos\ x, \quad f'''(0) = 0.$$

In general

$$f^{(n)}(x) = \begin{cases} sin\ x & n = 4k, \quad k = 0, 1, \ldots; \\ cos\ x & n = 4k + 1, \quad k = 0.1, \ldots; \\ -sin\ x & n = 4k + 2, \quad k = 0, 1, \ldots; \\ -cos\ x & n = 4k + 3, \quad k = 0, 1, \ldots; \end{cases}$$

and

$$f^{(n)}(0) = (sin\ 0)^{(n)} = \begin{cases} (-1)^k & n = 2k + 1, \quad k = 0, 1, \ldots; \\ 0 & n = 2k, \quad k = 0, 1, \ldots; \end{cases}$$

Thus, Taylor polynomial for $sin\ x$ is

$$TL_{2n+1}(x) = x - \frac{x^3}{3!} + \frac{x^5}{5!} - \frac{x^7}{7!} + \cdots + (-1)^{n+1}\frac{x^{2n+1}}{(2n + 1)!},$$

where the Lagrange's remainder

$$R_{2n+2}(\xi_x) = \frac{(sin\ \xi_x)^{(2n+2)}}{(2n + 2)!}x^{2n+2}.$$

One can obtain the Taylor's polynomial $TL_9(x)$ using the instruction

```
Normal[Series[Sin[x],{x,0,9}]]
```

In order to get correct three decimal places, we should consider accuracy $\epsilon = 0.0005$, and to choose a smallest n for which the remainder $R_{n+1}(\xi_x)$ satisfies the following inequality

$$\left| \frac{(sin\xi_x)^{(2n+2)}}{(2n+2)!} x^{2n+2} \right| \le \epsilon.$$

Obviously, the above inequality holds if

$$\frac{1}{(2n+2)!} \left(\frac{\pi}{2}\right)^{2n+2} \le 0.0005.$$

Hence $n = 4$, so that the Taylor polynomial

$$TL_9(x) = x - \frac{x^3}{3!} + \frac{x^5}{5!} - \frac{x^7}{7!} + \frac{x^9}{9!}$$

approximates $sin\, x$ in the interval $[0, \frac{\pi}{2}]$ with accuracy upto three decimal places.

Solution of (b), (ii) For the function $\cos x$, we find

$$f^{(n)}(x) = (\cos x)^{(n)} = \begin{cases} \cos x & \text{for } n = 4k, \ k = 0, 1, \ldots; \\ -\sin x & \text{for } n = 4k+1, \ k = 0, 1, \ldots; \\ -\cos x & \text{for } n = 4k+2, \ k = 0, 1, \ldots; \\ \sin x & \text{for } n = 4k+3, \ k = 0, 1, \ldots; \end{cases}$$

and

$$cos^{(n)}0 = \begin{cases} (-1)^k & \text{for } n = 2k, \ k = 0, 1, \ldots; \\ 0 & \text{for } n = 2k+1, \ k = 0, 1, \ldots; \end{cases}$$

Thus, Taylor polynomial

$$TL_{2n}(x) = 1 - \frac{x^2}{2!} + \frac{x^4}{4!} + \cdots + (-1)^n \frac{x^{2n}}{2n!},$$

where the remainder

$$R_{2n+1}(\xi_x) = (-1)^{n+1} \frac{(\cos\xi_x)^{(2n+1)}}{(2n+1)!} x^{2n+1}.$$

In order to get accuracy of three decimal places, we choose $n = 4$ and the polynomial

$$TL_8(x) = 1 - \frac{x^2}{2!} + \frac{x^4}{4!} - \frac{x^6}{6!} + \frac{x^8}{8!}.$$

Then, for $n = 4$, we have the following remainder estimate

$$| R_{2n+1}(\xi_x) | \;\; = \;\; | (-1)^{n+1} \frac{(cos\ \xi_x)^{(2n+1)}}{(2n+1)!} |$$

$$\leq \;\; \frac{1}{(2n+1)!}(\frac{\pi}{2})^{(2n+1)} \leq 0.0005.$$

Also, we can obtain the Taylor's polynomial $TL_8(x)$ by the `Mathematica` instruction

```
Normal[Series[Cos[x],{x,0,8}]]
```

4.2   Bernstein Polynomials and Weierstrass Theorem

In the previous section, it has been shown that an $(n + 1)$ times continuously differentiable function f can be approximated by its Taylor polynomial in a neighbourhood of a point x_0, provided that the remainder $R_{n+1}(\xi_x)$ tends to zero when $n \to \infty$. A continuous function f (not necessarily differentiable) on the interval $[a, b]$ can also be approximated by a polynomial

$$p_n(x) = a_0 + a_1 x + a_2 x^2 + \cdots + a_n x^n,$$

with an arbitrary accuracy ϵ. This has been established by Weierstrass in the following theorem:

Theorem 4.2 *(Weierstrass Theorem.) If f is a continuous function on a finite and closed interval $[a, b]$, then for every $\epsilon > 0$ there exists a polynomial $p_n(x)$ such that*

$$| f(x) - p_n(x) | < \epsilon \quad for \;\; all \quad x \in [a, b]$$

Proof.   There are a few proofs of Weierstrass approximation theorem. Some of them have constructive character and they lead to an explicit form of the polynomial $p_n(x)$. Such proof due to Bernstein (1912) is presented below. Let us first consider the case when the interval $[a, b] = [0, 1]$. Then, we shall show that the sequence $B_n(x)$ of Bernstein's polynomials

$$B_n(x)= \sum_{k=0}^{n} f(\frac{k}{n})x^k(1 - x)^{n-k}, \quad \text{for} \quad n = 0, 1, \ldots;$$

is uniformly convergent to the function $f(x)$ on the interval $[0, 1]$. The following identities hold:

(i) $\sum_{k=0}^{n} C_k^n x^k (1 - x)^{n-k} = 1,$

(ii) $\sum_{k=0}^{n} \frac{k}{n} C_k^n x^k (1 - x)^{n-k} = x,$ (4.2)

(iii) $\sum_{k=0}^{n} \frac{k^2}{n^2} C_k^n x^k (1 - x)^{n-k} = (1 - \frac{1}{n})x^2 + \frac{1}{n}x,$

where $C_k^n = \dfrac{n!}{k!(n - k)!}$ is Newton's coefficient.

We can easily get the inequality (i) setting $u = x$ and $v = 1 - x$ in Newton's binomial

$$(u + v)^n = \sum_{k=0}^{n} C_k^n u^k v^{n-k}.$$

Also, we can obtained identities (ii) and (iii) by differentiating Newton's binomial once and twice, and then substituting $u = x$ and $v = 1 - x$.

The identities (i), (ii) and (iii) yield the following equality:

(iv) $\sum_{k=0}^{n} C_k^n (\frac{k}{n} - x)^2 x^k (1 - x)^{n-k} = \dfrac{x(1 - x)}{n}.$

for all $x \in [0, 1]$.

Now, let us multiply both hand sides of (i) by $f(x)$ and then subtract $B_n(x)$. Then, we have

$$f(x) - B_n(x) = \sum_{k=0}^{n} [f(x) - f(\frac{k}{n})] C_k^n x^k (1-x)^{n-k}.$$

By assumption, $f(x)$ is a continuous function on the closed interval $[0, 1]$. Therefore, $f(x)$ is also a function uniformly continuous and a bounded on the interval $[0, 1]$, *i.e.*, for every $\epsilon > 0$, there exists $\delta_\epsilon > 0$ such that

$$\mid f(x_1) - f(x_2) \mid < \frac{\epsilon}{2} \quad \text{if} \quad \mid x_1 - x_2 \mid < \delta_\epsilon,$$

and there exists a constant M such that

$$\mid f(x) \mid \leq M \quad \text{for all} \quad x \in [0.1].$$

Let us divide the set of points

$$S = \{0, \frac{1}{n}, \frac{2}{n}, \ldots, \frac{n}{n}\}$$

into two disjoint subsets

$$S_1 = \{\frac{k}{n} : \mid \frac{k}{n} - x \mid < \delta_\epsilon\} \quad \text{and} \quad S_2 = \{\frac{k}{n} : \mid \frac{k}{n} - x \mid \geq \delta_\epsilon\}.$$

Then, we have

$$\mid f(x) - B_n(x) \mid = \mid \sum_{S_1} [f(x) - f(\tfrac{k}{n})] C_k^n x^k (1-x)^{n-k} \mid$$

$$< \tfrac{\epsilon}{2} \sum_{S_1} C_k^n x^k (1-x)^{n-k} \leq \tfrac{\epsilon}{2}$$

for all $\mid \frac{k}{n} - x \mid < \delta_\epsilon$.

By formulas (i), (ii) and (iii) and by the inequality

$$x(1-x) \leq \frac{1}{4}, \quad x \in [0, 1],$$

we have

$$\sum (x - \frac{k}{n})^2 C_k^n x^k (1-x)^{n-k} = \frac{x(1-x)}{n} \leq \frac{1}{4n}.$$

Hence

$$
\begin{aligned}
|f(x) - B_n(x)| \ &= |\sum_{S_2}[f(x) - f(\frac{k}{n})]C_k^n x^k (1-x)^{n-k}| \\
&\leq 2M \sum_{S_2} C_k^n x^k (1-x)^{n-k} \\
&\leq 2M \sum_{S_2} (\frac{k-nx}{k-nx})^2 C_k^n x^k (1-x)^{n-k} \\
&\leq 2M \frac{x(1-x)}{n} \leq \frac{M}{2n\delta_\epsilon^2}.
\end{aligned}
$$

Let us note that the inequality

$$
\frac{M}{2n\delta_\epsilon} < \frac{\epsilon}{2}.
$$

holds for sufficiently large n.

Finally, combining the above inequalities, we obtain

$$
| f(x) - B_n(x) | < \epsilon \quad \text{for all} \quad x \in [0,1].
$$

In the case when $[a,b] \neq [0,1]$, we may apply the following mapping:

$$
\bar{x} = \frac{1}{b-a}[x-a].
$$

to map the interval $[a,b]$ to the interval $[0,1]$.

The rate of convergence of Bernstein's polynomials $B_n(x)$, $n = 0, 1, \ldots$, to the function f depends on regularity of f. In principle more regular functions can be approximated by Bernstein's polynomials with higher rate of convergence.

Let us assume that f satisfies the Lipschitz condition, *i.e.*, there exists a constant L such that

$$
| f(x') - f(x'') | \leq L | x' - x'' |,
$$

for any $x', x'' \in [a,b]$.

Then, the rate of convergence of $B_n(x)$ to $f(x)$, when $n \to \infty$ is $O(\frac{1}{\sqrt{n}})$, *i.e.*,

$$
| f(x) - B_n(x) | \leq \frac{L}{2\sqrt{n}}
$$

for all $n = 1, 2, \ldots$;
Namely, we observe that

$$|f(x) - B_n(x)| \leq \sum_{k=0}^{n} |f(x) - f(\frac{k}{n})| C_k^n x^k (1-x)^{n-k} \text{ for } x \in [a, b].$$

By the Lipschitz condition

$$|f(x) - B_N(x)| \leq L \sum_{k=0}^{n} |x - \frac{k}{n}| C_k^n x^k (1-x)^{n-k}.$$

Then, by Buniakowsky inequality

$$\sum_{k=0}^{n} |x - \frac{k}{n}| C_k^n x^k (1-x)^{n-k} \leq$$

$$\leq \sqrt{\sum_{k=0}^{n} (x - \frac{k}{n})^2 C_k^n x^k (1-x)^{n-k}} \sqrt{\sum_{k=0}^{n} C_k^n x^k (1-x)^{n-k}}.$$

Since

$$\sum_{k=0}^{n} (x - \frac{k}{n})^2 C_k^n x^k (1-x)^{n-k} = \frac{x(1-x)}{n} \leq \frac{1}{4n},$$

we get the following estimation of the rate of convergence:

$$|f(x) - B_n(x)| \leq \frac{L}{2\sqrt{n}} \quad \text{for all} \quad x \in [0, 1].$$

Now, let us assume that function f is twice continuously differentiable in the interval $[a, b]$. Then, we shall show that the rate of convergence of Bernstein's polynomials $B_n(x)$, $n = 0, 1, \ldots$; to the function f is $O(\frac{1}{n})$. Namely, we observe that (*cf.* [2,3,7,9, 11])

$$
\begin{aligned}
f(x) - B_n(x) &= \sum_{k=0}^{n} [f(x) - f(\frac{k}{n})] C_k^n x^k (1-x)^{n-k} \\
&= \sum_{k=0}^{n} [f'(x)(x - \frac{k}{n}) + \frac{f''(x)}{2}(x - \frac{k}{n})^2 \\
&+ \frac{R_n}{n}] C_k^n x^k (1-x)^{n-k} = \frac{f''(x)}{2n} x(1-x) + \frac{R_n}{n},
\end{aligned}
$$

where R_n tends to zero when $n \to \infty$.

From the above formula, it follows that the rate of convergence cannot be better than $O(\frac{1}{n})$, except for the case when f is a linear function. Therefore, in practice, we use a more effective method of approximation than the method based on Weierstrass theorem.

4.3   Best Approximation Polynomials

Minimax Problem. From Weierstrass theorem, we know that every continuous function on a finite and closed interval $[a, b]$ can be uniformly approximated by a polynomial $P_n(x)$ with arbitrary accuracy $\epsilon > 0$. However, the rate of convergence $O(\frac{1}{\sqrt{n}})$ of the polynomials $P_n(x)$ to $f(x)$ when $n \to \infty$ is slow. More effective approximation of a continuous function f can be reached by optimal polynomials which we shall describe below.

Let

$$E(f, P_n) = \max_{a \le x \le b} |f(x) - P_n(x)|$$

be the error of uniform approximation of the function $f(x)$ by a polynomial $P_n(x)$ on the interval $[a, b]$. The polynomial $P_n(x)$ which minimizes the error $E(f, P_n)$ is the solution of the following "*minimax problem*":

Among all possible polynomials $P_n(x)$ of degree at most n, find one for which the error $E(f, P_n)$ attains its minimum, i.e..

$$E(f) = \inf_{P_n} E(f, P_n) = \inf_{P_n} \max_{a \le x \le b} |f(x) - P_n(x)|. \qquad (4.3)$$

The above "*minimax problem*" possesses a unique solution, *i.e.*, among all possible polynomials of degree at most n there exists only one which satisfies formula (4.3).

Proof of existence.   Let

$$P_n(x) = a_0 + a_1 x + a_2 x^2 + \cdots + a_n x^n$$

be a polynomial with real coefficients $a_0, a_1, \ldots, a_n$ of degree at most n and not identically equal to zero. Let us consider the following functions:

$$G(a_0, a_1, \ldots, a_n) = \max_{a \le x \le b} |a_0 + a_1 x + a_2 x^2 + \cdots + a_n x^n|,$$

$$H(a_0, a_1, \cdots, a_n) = \max_{a \le x \le b} |f(x) - a_0 - a_1 x - a_2 x^2 - \cdots - a_n x^n|.$$

The functions $G(a_0, a_1, \ldots, a_n)$ and $H(a_0, a_1, \ldots, a_n)$ depend on $(n+1)$ parameters $a_0, a_1, \ldots, a_n$ and they are continuous in the whole real space R^{n+1}. Therefore, $G(a_0, a_1, \ldots, a_n)$ and $H(a_0, a_1, \ldots, a_n)$ are also continuous on the closed unit ball K

$$a_0^2 + a_1^2 + \cdots + a_n^2 = 1.$$

By Weierstrass theorem, the function $G(a_0, a_1, \cdots, a_n)$ attains its positive minimum $\mu > 0$ on the ball K, i.e.,

$$G(a_0, a_1, \ldots, a_n) \ge \mu > 0 \quad \text{for } (a_0, a_1, \ldots, a_n) \in K.$$

This minimum μ must be positive, since

$$G(a_0, a_1, \ldots, a_n) = 0 \text{ if and only if } a_0 = a_1 = \cdots = a_n = 0,$$

and the polynomial $P_n(x)$ is not identically equal to zero. The function $H(a_0, a_1, \cdots, a_n)$ is non-negative and has non-negative minimum in the whole space R^{n+1}. Let

$$m = \inf_{R^{n+1}} H(a_0, a_1, \ldots, a_n) \ge 0.$$

Now, let us divide the space R^{n+1} into two sets as follows:

$$S_1 = \{(a_0, a_1, \ldots, a_n) \; : \; \sum_{i=0}^{n} a_i^2 \le r^2\},$$

$$S_2 = \{(a_0, a_1, \ldots, a_n) \; : \; \sum_{i=0}^{n} a_i^2 > r^2\},$$

where

$$r = \frac{1}{\mu}(1 + m + \max_{a \le x \le b} |f(x)|) > 0.$$

Then, we get the inequality

$$\sum_{i=0}^{n} a_i^2 = \lambda^2 > r^2$$

for any $(a_0, a_1, \ldots, a_n) \in S_2$, and

$$
\begin{aligned}
H(a_0, a_1, \ldots, a_n) &= \max_{a \leq x \leq b} |f(x) - \sum_{i=0}^{n} a_i x^i| \\
&\geq \max_{a \leq x \leq b} |\sum_{i=0}^{n} a_i x^i| - \max_{a \leq x \leq b} |f(x)| \\
&= \lambda \max_{a \leq x \leq b} |\sum_{i=0}^{n} \frac{a_i}{\lambda} x^i| - \max_{a \leq x \leq b} |f(x)| \\
&\geq \lambda \mu - \max_{a \leq x \leq b} |f(x)| \geq r\mu - \max_{a \leq x \leq b} |f(x)| \\
&\geq 1 + m
\end{aligned}
$$

for all $(a_0, a_1, \ldots, a_n) \in S_2$.

Hence, the function $H(a_0, a_1, \ldots, a_n)$ attains its minimum equal to m on the set S_1, (not on S_2), *i.e.*, there exists a point

$$(a_0^*, a_1^*, \ldots, a_n^*) \in S_1$$

such that

$$
\begin{aligned}
H(a_0^*, a_1^*, \ldots, a_n^*) &= \min_{S_1} H(a_0, a_1, \ldots, a_n) \\
&= \inf_{R^{n+1}} H(a_0, a_1, \ldots, a_n) = m.
\end{aligned}
$$

Hence

$$P_n(x) = a_0^* + a_1^* x + a_2^* x^2 + \cdots + a_n^* x^n$$

is the best approximation polynomial to the function $f(x)$ on the interval $[a, b]$.

Proof of Uniqueness. Let us assume that there are two different polynomials $P_n(x)$ and $Q_n(x)$ which approximate the function f on the interval $[a, b]$ with the minimum error $E(f, P_n)$. Then, we have

$$\max_{a \leq x \leq b} |f(x) - P_n(x)| = \max_{a \leq x \leq b} |f(x) - Q_n(x)| = m.$$

Let us consider the ball

$$B(f) = \{g \in C_0[a,b] \; : \; \max_{a \leq x \leq b} |f(x) - g(x)| \leq m\}$$

[2] The polynomials $P_n(x)$, $Q_n(x)$ and

$$R_n^{(q)}(x) = qP_n(x) + (1-q)Q_n(x), \qquad 0 \leq q \leq 1,$$

are members of the ball $B(f)$, i.e., $P_n, Q_n, R_n^{(q)} \in B(f)$
Also, we have

$$
\begin{aligned}
m & \leq & \max_{a \leq x \leq b} |f(x) - R_n^{(q)}(x)| \\
& = & \max_{a \leq x \leq b} |f(x) - qP_n(x) - (1-q)Q_n(x)| \\
& \leq & q \max_{a \leq x \leq b} |f(x) - P_n(x)| \\
& + & (1-q) \max_{a \leq x \leq b} |f(x) - Q_n(x)| = m.
\end{aligned}
$$

Hence

$$\max_{a \leq x \leq b} |f(x) - R_n^{(q)}(x)| = m, \qquad\qquad (4.4)$$

and therefore $R_n^{(q)} \in B(f)$.
But, for $P_n(x) \neq Q(x)$ and $0 < q < 1$, polynomial $R_n^{(q)}(x)$
lies within the ball $B(f)$, i.e.,

$$\max_{a \leq x \leq b} |f(x) - R_n^{(q)}(x)| < m$$

This inequality contradicts to (4.4). Thus, $P_n(x) = Q_n(x)$
for all $x \in [a,b]$.
Chebyshev equi-oscillation theorem.   In general, most
algorithms to find the best approximation polynomial $P_n(x)$
for a given continuous function $f(x)$ on a closed interval
$[a,b]$ are based on the following Chebyshev equi-oscillation
theorem:

---

[2]$C_0[a,b]$ is the set of all continuous functions on the closed interval $[a,b]$

Theorem 4.3 *Let f be a continuous function on a finite and closed interval $[a, b]$ and let P_n be a polynomial with the oscillation*

$$L = \max_{a \leq x \leq b} |f(x) - P_n(x)|.$$

Then, $P_n(x)$ is the best approximating polynomial to $f(x)$ on the interval $[a, b]$ if and only if there exist at least $n + 2$ points

$$a \leq x_0 < x_1 < \cdots < x_{n+1} \leq b$$

such that

$$|f(x_i) - P_n(x_i)| = L \quad \text{for} \quad i = 0, 1, \ldots, n + 1 \qquad (4.5)$$

and

$$f(x_i) - P_n(x_i) = -[f(x_{i+1}) - P_n(x_{i+1})] \quad \text{for} \quad i = 0, 1, \ldots, n. \tag{4.6}$$

The points x_i, $i = 0, 1, \ldots, n + 1$ which satisfy the above conditions (4.5) and (4.6) are called Chebyshev knots.

Proof.   We shall first prove that (4.5) and (4.6) are necessary conditions for $P_n(x)$ to be the best approximating polynomial to the function $f(x)$.

Necessity.   Let us assume that $P_n(x)$ is the best approximating polynomial to the function $f(x)$ on the interval $[a, b]$ and, on the contrary, that the conditions (4.5) and (4.6) are satisfied for certain number of Chebyshev knots $s < n + 2$. The number s must be greater than one *i.e.*, $s \geq 2$. Indeed, if $s = 1$ then the error function

$$E(x) = f(x) - P_n(x), \quad x \in [a, b]$$

attains its maximum at x_0 *i.e.*,

$$E(x_0) = f(x_0) - P_n(x_0) = L$$

or $E(x)$ attains its minimum at x_0 *i.e.*,

$$E(x_0) = f(x_0) - P_n(x_0) = -L.$$

Therefore

$$-L \leq E(x) \leq L \quad \text{for all} \quad x \in [a, b].$$

Let $E(x_0) = L$. As a continuous function on the closed interval $[a, b]$, $E(x)$ attaints its minimum at a point $x_1 \in [a, b]$, and for $s = 1$

$$\min_{a \leq x \leq b} E(x) = E(x_1) > -L.$$

Then, the polynomial

$$\overline{P}_n(x) = P_n(x) + \frac{1}{2}[E(x_1) + L]$$

approximates the function $f(x)$ better than the polynomial $P_n(x)$ does. Indeed, we have

$$\overline{E}(x) = f(x) - \overline{P}_n(x) = E(x) - \frac{1}{2}[E(x_1) + L].$$

and

$$-L < \overline{E}(x) < L \quad \text{for all} \quad x \in [a, b].$$

Now, let $E(x_0) = -L$. Then $E(x)$ attains its maximum at x_1, and for $s = 1$.

$$\max_{a \leq x \leq b} E(x) = E(x_1) < L.$$

In this case, the polynomial

$$\overline{P}_n(x) = P_n(x) + \frac{1}{2}[L - E(x_1)]$$

approximates the function $f(x)$ better than the polynomial $P_n(x)$ does. Indeed, we have

$$\overline{E}(x) = f(x) - \overline{P}_n(x) = E(x) + \frac{1}{2}[L - E(x_1)],$$

and

$$-L < -\frac{1}{2}L - \frac{1}{2}E(x_1) \leq E(x) + \frac{1}{2}[L - E(x_1)] \leq \frac{1}{2}E(x_1) + \frac{1}{2}L < L.$$

Hence

$$-L < \overline{E}(x) < L \quad \text{for all } x \in [a, b].$$

We may illustrate the above relations, on the following figures (**see Fig 4.1, Fig 4.2**):

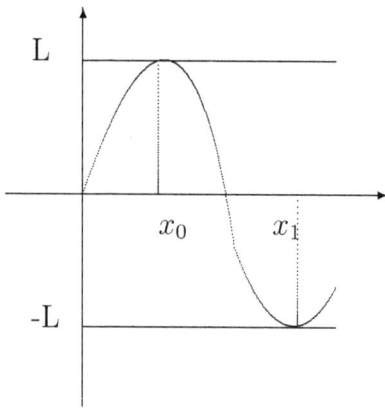   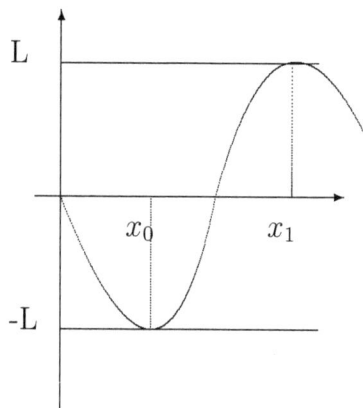

Fig. 4.1   **Fig. 4.2**

Thus, if $P_n(x)$ is the best approximating polynomial to $f(x)$, then the number of knots $s \geq 2$.

Thus, if $P_n(x)$ is the best approximating polynomial to $f(x)$, then the number of knots $s \geq 2$.

Now, let assume that $s \leq n + 1$. In this case, we shall also find a polynomial $\overline{P}_n(x)$ which is better Chebyshev approximation to $f(x)$ than the best polynomial $P_n(x)$. Namely, let us divide the interval $[a, b]$ into subintervals $I_1, I_2, \ldots, I_N$ each of the same length δ. Since $E(x)$ is a uniformly continuous function on the interval $[a, b]$, we get

$$|E(x^{'}) - E(x^{''})| \leq \max_{a \leq x \leq b} |E(x)|$$

for all $x', x'' \in [a, b]$ such that $|x' - x''| < \delta$. Let $J_1, J_2, \ldots, J_s$, be the subintervals (chosen from $I's$ intervals) on which

$$\max_{x \in J_i} |E(x)| = L, \quad i = 1, 2, \ldots, s.$$

Let us choose δ so small to be a non-empty "gap" between two following subintervals, J_i and J_{i+1}, $i = 1, 2, \ldots, s - 1$ and such that on each subinterval J_i, $i = 1, 2 \ldots, s$, $E(x)$ has a constant sign (**see Fig 4.3**). Such choice of δ is possible, since $E(x)$ is uniformly continuous function on interval $[a, b]$.

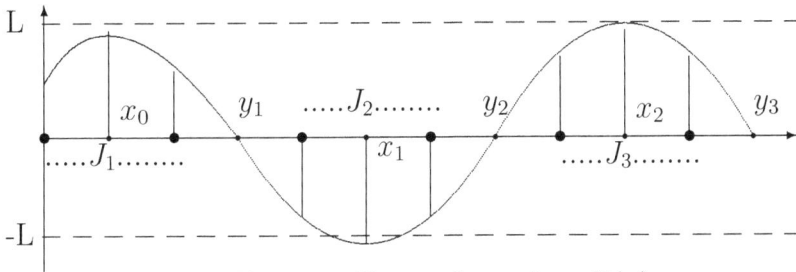

Fig 4.3. Error function $E(x)$.

Now, let us choose $y_1, y_2, \ldots, y_{s-1}$ from the "gaps", so that y_i lies between J_i and J_{i+1}, for $i = 1, 2, \ldots, s-1$. Under the assumption $s \leq n + 1$, we shall show that the polynomial

$$\overline{P}_n(x) = P_n(x) + \epsilon Q_{s-1}(x),$$

where

$$Q_{s-1}(x) = (x - y_1)(x - y_2) \ldots (x - y_{s-1}),$$

is better Chebyshev approximation to the function $f(x)$ than the polynomial $P_n(x)$, if $\epsilon > 0$ is sufficiently small. Indeed, we have

$$\overline{E}(x) = f(x) - \overline{P}_n(x) = E(x) - \epsilon Q_{s-1}(x).$$

The polynomial $Q_{s-1}(x) > 0$ for $x \in J_1$ and $Q_{s-1}(x)$ changes its sign only at y_i, $i = 1, 2, \ldots, s - 1$. If $E(x) > 0$ for $x \in J_1$ then $Q_{s-1}(x)$ and $E(x)$ have the same signs on J_i, $\quad i =$

$1, 2, \ldots, s$. Clearly, for sufficiently small $\epsilon > 0$ the following inequality holds:

$$\epsilon|Q_{s-1}(x)| < |E(x)| \quad \text{for} \quad x \in [a, b]$$

and

$$\max_{x \in J_i} |\overline{E}(x)| < \max_{x \in J_i} |E(x)| \quad \text{for} \quad i = 1, 2, \ldots, s.$$

Hence

$$-L < \overline{E}(x) < L \quad \text{for } x \in J_i, \quad i = 1, 2, \ldots, s. \qquad (4.7)$$

To complete the proof, we must show that inequality (4.7) is satisfied for all $x \in [a, b]$.
Let

$$\Omega = [a, b] - J_1 \cup J_2 \cup \cdots \cup J_s.$$

Then

$$\max_{\Omega} |\overline{E}(x)| \leq \max_{\Omega} |E(x)| + \epsilon \max_{\Omega} |Q_s(x)|$$

and

$$\max_{\Omega} |E(x)| < \max_{a \leq x \leq a} |E(x)|.$$

Now, let us choose $\epsilon > 0$ so small to be

$$\epsilon|Q_s(x)| < \max_{a \leq x \leq b} |E(x)| - \max_{\Omega} |E(x)| \quad \text{for} \quad x \in [a, b].$$

Then, by the above inequalities

$$\max_{\Omega} |\overline{E}(x)| < \max_{a \leq x \leq b} |E(x)| = L.$$

Hence, by (4.7)

$$-L < \overline{E}(x) < L \text{ for all } x \in [a, b].$$

This means that the polynomial $\overline{P}_n(x)$ approximates the function $f(x)$ better than $P_n(x)$ does, provided that $s \leq n + 1$. Thus, $s \geq n + 2$, if $P_n(x)$ is the best approximating polynomial to the function $f(x)$ on the interval $[a, b]$.
Sufficiency. Let us assume that polynomial $P_n(x)$ satisfies

conditions (4.5) and (4.6) *i.e.*, there exist at least $n+2$ points

$$a \le x_0 < x_1 < \ldots < x_{n+1} \le b$$

such that

$$|f(x_i) - P_n(x_i)| = L, \quad i = 0, 1, 2, \ldots, n+1,$$

and

$$f(x_i) - P_n(x_i) = -[f(x_{i+1}) - P_n(x_{i+1})], \quad i = 0, 1, \ldots, n.$$

Also, for contrary, we assume that $P_n(x)$ is not the best Chebyshev approximating to $f(x)$ on the interval $[a, b]$, and $\overline{P}_n(x)$ is the one. Now, let us consider the following polynomial:

$$\Psi_n(x) = \overline{P}_n(x) - P_n(x).$$

Then, we have

$$\Psi_n(x) = [f(x) - P_n(x)] - [f(x) - \overline{P}_n(x)]. \quad (4.8)$$

From the assumptions (4.5) and (4.6), it follows that

$$f(x_i) - P_n(x_i) = \pm L, \quad i = 0, 1, \ldots, n+1.$$

$\overline{P}_n(x)$ is the best Chebyshev approximation of $f(x)$, therefore

$$|f(x) - \overline{P}_n(x)| = |\overline{E}(x)| = \overline{L} < L, \quad a \le x \le b.$$

From the above, by (4.8), we conclude that polynomial $\Psi_n(x)$ of degree at most n has a root in each subinterval (x_i, x_{i+1}) , $i = 0, 1, \ldots, n$. Thus, $\Psi_n(x)$ possesses at least $n+1$ roots. This is possible when $\Psi_n(x) = 0$ for all $x \in [a, b]$. Therefore, $P_n(x) = \overline{P}_n(x)$ for all $x \in [a, b]$, and $P_n(x)$ must be the best Chebyshev approximation to $f(x)$ on interval $[a, b]$.

An application of the Chebyshev theorem. A special interest is paid to approximate the function $f(x) = 0$ for all $x \in [a, b]$ by a polynomial $P_n(x)$ with the leading coefficient at x^n equal to one. We have used this approximation of a *"zero"* function to determine an optimal interpolating polynomial to a function $f(x)$. So, we have used the following thesis:

Among all polynomials of degree n with the leading coefficient at x^n equal to one the polynomial

$$\overline{T}_n(x) = \frac{1}{2^{n-1}} \cos\left(n \ arc \ cos \ x\right), \quad x \in [-1, 1]$$

is the best approximation of the "zero" function $f(x) = 0$ in the interval $[-1, 1]$.

Now, we can prove this thesis using the equi-oscillation Chebyshev theorem. Namely, let us note that the polynomial $\overline{T}_n(x)$ attains its maximum modulus on the interval $[-1, 1]$ equal to $\frac{1}{2^{n-1}}$ at the following $n + 1$ points:

$$x_k = \cos\frac{k\pi}{n}, \quad k = 0, 1, \dots, n.$$

Indeed, the derivative

$$\frac{d\overline{T}_n(x)}{dx} = \frac{n \ sin \ (n \ arccos \ x)}{\sqrt{1 - x^2}} = 0, \quad x \in (-1, 1)$$

if

$$n \ arc \ cos \ x_k = k\pi.$$

Then

$$x_k = \cos\frac{k\pi}{n}, \quad k = 1, 2, \dots, n - 1.$$

Also, $\overline{T}_n(x)$ attains its maximum modulus $\dfrac{1}{2^{n-1}}$ at the end of the interval $[-1, 1]$.

Hence, we have

$$\overline{T}_n(x_k) = \frac{1}{2^{n-1}} \cos n \ (arccos(\cos\frac{k\pi}{n})) = \frac{(-1)^k}{2^{n-1}}$$

for $k = 0, 1, \dots, n$.

Therefore, the function $\overline{T}_n(x)$ satisfies the assumptions (4.5) and (4.6) of the equi-oscillation theorem, and $\overline{T}_n(x)$ is the best Chebyshev approximation of the "*zero*" function on interval $[-1, 1]$.

Example 4.4 *Find the best approximation of the polynomial*

$$f(x) = x^n, \quad -1 \leq x \leq 1,$$

by a polynomial of degree at most $n - 1$ with the leading coefficient equal to one at x^{n-1}.

This kind of approximation maybe interesting when we need to minimize the number of terms in the following polynomial:

$$f(x) = a_0 + a_1 x + a_2 x^2 + \cdots + a_n x^n$$

with minimum loss of accuracy.

In order to determine the best approximating polynomial $P_{n-1}(x)$ to x^n, we consider the following error function

$$E(x) = x^n - P_{n-1}(x), \quad -1 \leq x \leq 1.$$

This error function attains its minimum oscillation if

$$x^n - P_{n-1}(x) = \overline{T}_n(x), \quad -1 \leq x \leq 1.$$

Hence

$$P_{n-1}(x) = x^n - \overline{T}_n(x), \quad -1 \leq x \leq 1.$$

4.4   Chebyshev Series

We may use the Chebyshev polynomials to approximate a continuous function by a partial sum of the following Chebyshev series:

$$\frac{a_0}{2} + \sum_{k=1}^{\infty} a_k T_k(x) = \frac{a_0}{2} T_0(x) + a_1 T_1(x) + \cdots + a_n T_n(x) + \cdots$$

for $x \in [-1, 1]$, where $T_k(x) = cos\ (k\ arccos\ x)$ and the coefficients

$$a_k = \frac{2}{\pi} \int_{-1}^{1} \frac{f(x)T_k(x)}{\sqrt{1-x^2}}, \quad k = 0, 1, \ldots,$$

In fact, the above Chebyshev series reduces to Fourier's series of cosines of the function $f(cos\ \Theta)$, where $x = cos\ \Theta$, $\Theta \in [0, \pi]$.

Indeed, for $x = cos\ \Theta$, we have $dx = -sin\ \Theta\ d\Theta$ and

$$a_k = \frac{2}{\pi} \int_{-1}^{1} \frac{f(x)T_k(x)}{\sqrt{1-x^2}} dx =$$

$$= \frac{2}{\pi} \int_{0}^{\pi} f(cos\ \Theta)\ cos\ k\Theta\ d\Theta.$$

On the other hand, the Fourier's series of cosines of $f(cos\ \Theta)$ takes the following form:

$$\frac{A_0}{2} + \sum_{k=1}^{\infty} A_k\ cos\ k\Theta + B_k\ sin\ k\Theta,$$

where

$$A_k = \frac{1}{\pi} \int_{-\pi}^{\pi} f(cos\ \Theta) cos\ k\Theta\ d\Theta, \quad k = 0, 1, \ldots;$$

$$B_k = \frac{1}{\pi} \int_{-\pi}^{\pi} f(cos\ \Theta) sin\ \Theta\ d\Theta, \quad k = 1, 2, \ldots;$$

Since $f(\cos \Theta) \cos \Theta$ is an even function and $f(\cos \Theta) \sin \Theta$ is an odd function, we get

$$A_k = \frac{1}{\pi} \int_{-\pi}^{\pi} f(\cos \Theta) \cos \Theta\ d\Theta = \frac{2}{\pi} \int_{0}^{\pi} f(\cos \Theta) \cos k\Theta\ d\Theta = a_k$$

and

$$B_k = \frac{1}{\pi} \int_{-\pi}^{\pi} f(\cos \Theta) \sin\ n\ \Theta\ d\Theta = 0.$$

So, the Chebyshev series of $f(x)$ is the Fourier's series of $f(cos\ \Theta)$, where $x = cos\ \Theta$.

Therefore, the Chebyshev series converges to $f(x)$ when

Fourier's series of cosines converges to $f(\cos\Theta)$. Then, for continuous functions

$$f(x) = a_0T_0(x) + a_1T_1(x) + \cdots + a_nT_n(x) + \cdots; \quad x \in [-1,1].$$

To obtain Chebyshev's series for a continuous function $f(x)$ in the interval $[a,b]$, we can proceed as follows:

1. (a) i. We apply the linear mapping

$$x = \frac{b-a}{2}(t+1) + a, \quad t \in [-1,1].$$

to transform interval $[-1,1]$ on interval $[a,b]$.

ii. Then, we find Chebyshev's series for the function

$$F(t) = f(\frac{b-a}{2}(t+1) + a),$$

given in the interval $[-1,1]$,

iii. The Chebyshev series of $F(t)$ is:

$$F(t) = \frac{a_0}{2} + \sum_{k=1}^{\infty} a_kT_k(t), \quad t \in [-1,1],$$

where the coefficients

$$a_k = \frac{2}{\pi}\int_0^{\pi} F(\cos\Theta)\cos k\Theta \, d\Theta, \quad k = 0,1,\ldots,$$

iv. We get the Chebyshev's series in terms of x , by the substitution

$$t = 2\frac{x-a}{b-a} - 1, \quad a \le x \le b,$$

so that the series is:

$$f(x) = \frac{a_0}{2} + \sum_{k=1}^{\infty} a_kT_k(2\frac{x-a}{b-a} - 1), \quad a \le x \le b.$$

Example 4.5 *Determine the partial sum $s_4(f)$, of the Chebyshev's series for the function*

$$f(x) = e^{-x^2}, \quad -1 \le x \le 1.$$

We shall find the partial sum of the series

$$e^{-x^2} = \frac{a_0}{2} + \sum_{k=1}^{\infty} a_k T_k(x), \quad -1 \leq x \leq 1,$$

using the following `Mathematica` program: `czebyszew1`

```
f[x_]:=Exp[-x^2];
ak[f_,n_]:=With[{coeff=Table[
NIntegrate[f[x]*ChebyshevT[i,x]/Sqrt[1-x^2],
        {x,-1,1}],{i,0,n}]},
N[ReplacePart[coeff,coeff[[1]]/2,1]*2/Pi]];
czebyszew1[n_]:=ak[f,n] Table[ChebyshevT[i,x],
        {i,0,n}]//Expand//Chop
```

By the instruction `czebyszew1[4]`, we obtain

$$s_4(f) = 0.996581 - 0.935316x^2 + 0.309633x^4.$$

Example 4.6 *Determine the partial sum $s_5(f)$ of the Chebyshev's series for the function*

$$f(x) = \sqrt{1 + x^2}, \quad 0 \leq x \leq 2.$$

Using the linear mapping $x = t + 1$, we transform the interval $[0, 2]$ onto interval $[-1, 1]$, so that, we consider the Chebyshev's series for the function

$$F(t) = \sqrt{1 + (1 + t)^2}, \quad -1 \leq t \leq 1.$$

By the program `czebyszew2[5]`,

```
f[x_]:=Sqrt[1+(1+x)^2];
ak[f_,n_]:=With[{coeff=Table[
        NIntegrate[f[Cos[x]]*Cos[i*x],
        {x,0,Pi}],{i,0,n}]},
    N[ReplacePart[coeff,coeff[[1]]/2,1]*2/Pi]];

czebyszew2[n_]:=ak[f,n].Table[ChebyshevT[i,x],
    {i,0,n}]//Expand//Chop
```

we find the partial sum

$$s_5(F) = 1.41399 + 0.707595t + 0.180681t^2 - 0.0923949t^3 + 0.0235995t^4 + 0.00277143t^5.$$

Coming back to the variable x, we have

$$F(t) = F(x-1) \approx s_5(f) = 1.41399 + 0.707595(x-1)$$
$$+ 0.180681(x-1)^2 - 0.0923949(x-1)^3$$
$$+ 0.0235995(x-1)^4 + 0.00277143(x-1)^5.$$

Hence, after simplification

$$s_5(f) = 1.0003 - 0.0114927x + 0.571748x^2 - 0.159078x^3$$
$$+ 0.00974232x^4 + 0.00277143x^5.$$

4.5   Exercises

Question 4.1 *Find the range of x for which*

　1. *(a)*

$$\sin x \approx x$$

　　(b)

$$\sin x \approx x - \frac{1}{6}x^3$$

with accuracy $\epsilon = 0.0001.$

Question 4.2 *Find the range of x for which*

　1. *(a)*

$$\cos x \approx 1 - \frac{1}{2}x^2$$

　　(b)

$$\cos x \approx 1 - \frac{1}{2}x^2 + \frac{1}{24}x^4$$

with accuracy $\epsilon = 0.0001.$

Question 4.3 *Let*

$$f(x) = \sqrt{1 + x^2}, \quad 0 \le x \le 1.$$

Find a linear function $y = ax + b$ which is the best approximate of $f(x)$

Question 4.4 *Consider the following approximation of $\sin x$:*

$$\sin x \approx p_5(x)$$

where

$$p_5(x) = x - \frac{1}{6}x^3 + \frac{1}{120}x^5.$$

Express $p_5(x)$ in terms of Chebyshev polynomials. Estimate the error $\sin x - p_5(x)$.

Question 4.5 *Use the* `Mathematica` *programs* `czebyszew1` *and* `czebyszew2` *to determine the partial sum $s_6(f)$ for the function $f(x) = ArcTan x$ in the intervals (i) $[-1, 1]$ and (ii) $[0, 4]$.*

Send Orders for Reprints to reprints@benthamscience.net

Lecture Notes in Numerical Analysis with Mathematica, 2014, 133-156   **133**

CHAPTER 5

Introduction to the Least Squares Analysis

Abstract

This chapter is designed for the least squares method . It begins with the least squares method to determine a polynomial of degree n well fitted to m discrete points $(x_i, y_i), i = 1, 2, ..., m$, when $n \leq m$. In the simplest form, the line of regression through m points is determined. The algorithm to find well fitted function to given discrete data is also presented. The Mathematica modules are designed and example illustrating the method are provided . The chapter ends with a set of questions.

Keywords: The least squares method, Regressions.

5.1   Introduction

In order to approximate the following discrete data:

x_0	x_1	x_2	x_3	$\cdots$	$\cdots$	$\cdots$	x_m
y_0	y_1	y_2	y_3	$\cdots$	$\cdots$	$\cdots$	y_m

we can use interpolating polynomial for a relatively small m. For large number of points, an interpolating polynomial

Krystyna STYŠ & Tadeusz STYŠ

is usually unstable and costly in terms of arithmetic operations. For large m, the least squares method is used to approximate discrete data points.

By least squares method, we determine a polynomial

$$P_n(x) = a_0^* + a_1^* x + a_2^* x^2 + \cdots + a_n^* x^n,$$

for which the function

$$
\begin{aligned}
D(a_0, a_1, \ldots, a_m) &= [y_0 - P_n(x_0)]^2 + [y_1 - P_n(x_1)]^2 \\
&+ [y_2 - P_n(x_2)]^2 + \cdots + [y_m - P_n(x_m)]^2
\end{aligned}
$$

attains its minimum at $a_0^*, a_1^*, \ldots, a_n^*$ in the whole real space R^{n+1}.

Let us note that $D(a_0, a_1, \ldots, a_n)$ is a non-negative quadratic function in the whole real space R^{n+1}. Therefore, this function attains its minimum if and only if

$$\frac{\partial D(a_0, a_1, a_2, \ldots, a_n)}{\partial a_k} = 0, \quad k = 0, 1, \ldots, n.$$

On the other hand

$$
\begin{aligned}
\frac{\partial D(a_0, a_1, a_2, \ldots, a_n)}{\partial a_k} &= \\
&= -2 \sum_{i=0}^{m} [y_i - a_0 - a_1 x_i - a_2 x_i^2 - \cdots - a_n x_i^n] x_i^k \\
&= -2 \sum_{i=0}^{m} x_i^k y_i + 2a_0 \sum_{i=0}^{m} + 2a_1 \sum_{i=0}^{m} x_i^{k+1} + 2a_2 \sum_{i=0}^{m} x_i^{k+2} + \cdots \\
&= 2a_n \sum_{i=0}^{m} x_i^{k+n} = 0.
\end{aligned}
$$

Hence, we obtain the following system of linear equations:

$$a_0 \sum_{i=0}^{m} x_i^k + a_1 \sum_{i=0}^{m} x_i^{k+1} + a_2 \sum_{i=0}^{m} x_i^{k+2} + \cdots + a_n \sum_{i=0}^{m} x_i^{k+n} = \sum_{i=0}^{m} x_i^k y_i$$

$$(5.1)$$

for $k = 0, 1, \ldots, n$.
Let

$$S_0 = m + 1, \quad S_k = \sum_{i=0}^{m} x_i^k,$$

$$V_0 = \sum_{i=0}^{m} y_i, \quad V_k = \sum_{i=0}^{m} x_i^k y_i,$$

for $k = 1, 2, \ldots, n$.
Let us write the system of linear equations (5.1) in the following form:

$$S_0 a_0 + S_1 a_1 + S_2 a_2 + \cdots + S_n a_n = V_0$$

$$S_1 a_0 + S_2 a_2 + S_3 a_3 + \cdots + S_n a_n = V_1$$

$$\ldots\ldots\ldots\ldots\ldots\ldots\ldots\ldots\ldots\ldots\ldots\ldots$$

$$\ldots\ldots\ldots\ldots\ldots\ldots\ldots\ldots\ldots\ldots\ldots\ldots$$

$$S_n a_0 + S_{n+1} a_1 + S_{n+2} a_2 + \cdots + S_{2n} a_{2n} = V_n$$

(5.2)

The system of equations (5.2) has a unique solution

$$a_0^*, a_1^*, a_2^*, \ldots, a_n^*$$

at which the function $D(a_0, a_1, a_2, \ldots, a_n)$ attains its minimum equal to
$D(a_0^*, a_1^*, a_2^*, \ldots, a_n^*)$. If $m = n$, then $D(a_0^*, a_1^*, a_2^*, \ldots, a_n^*) = 0$, and

$$P_n(x) = a_0^* + a_1^* x + a_2^* x^2 + \cdots + a_n^* x^n,$$

is the interpolating polynomial that satisfies the conditions

$$P_n(x_k) = y_k, \quad k = 0, 1, 2, \ldots, m.$$

In general, the number of points $m > n$, and then

$$D(a_0^*, a_1^*, \ldots, a_n^*) > 0,$$

so that $P_n(x)$ is not an interpolating polynomial. In such a case, we estimate the root mean square error using the following formula:

$$E = \sqrt{\frac{D(a_0^*, a_1^*, a_2^*, \ldots, a_n^*)}{m + 1}}.$$

Example 5.1 *Approximate the following discrete data by a least squares polynomial of degree two.*

0	1	2	3	4	5
1	−1	0	2	−1	0

Determine the error of approximation.

Solution. By the formulas (5.1), when $m = 5$ and $n = 2$, we find

$$S_0 = m + 1 = 6, \quad S_1 = 15, \quad S_2 = 55, \quad S_3 = 225,$$

$$S_4 = 979, \quad V_0 = 1, \quad V_1 = 1, \quad V_2 = 1.$$

Hence, we obtain the following system of linear equations:

$$6a_0 + 15a_1 + 55a_2 = 1,$$

$$15a_0 + 55a_1 + 225a_2 = 1,$$

$$55a_0 + 225a_1 + 979a_2 = 1.$$

Solving the above system of equations, we find the coefficients

$$a_0^* = 0.3214286, \quad a_1^* = 0.003571429, \quad a_2^* = -0.01785714.$$

and the polynomial

$$P_2(x) = 0.3214286 + 0.003571429x - 0.01785714x^2$$

that approximates discrete data with the root mean square error

$$E = \sqrt{\frac{D(a_0^*, a_1^*, a_2^*)}{m+1}}$$

$$= \frac{1}{\sqrt{6}}\{(1 - P_2(0))^2 + (-1 - P_2(1))^2$$

$$+ (0 - P_2(2))^2 + (2 - P_2(3))^2 + (-1 - P_2(4))^2$$

$$+ (0 - P_2(5))^2\}^{\frac{1}{2}} = 1.4469.$$

To obtain the quadratic polynomial $P_2(x)$, we execute the following commands (*cf.* [28]):

```
data={{0,1},{1,-1},{2,0},{3,2},{4,-1},{5,0}};
Fit[data,{1,x,x^2},x]
```

5.2   Line of Regression.

Special attention is paid to approximate discrete data by the linear function

$$P_1(x) = a_0^* + a_1^* x.$$

The graph of the linear function $P_1(x)$ which is the best fitted line to discrete data in the sense of least squares method is called *the line of regression.*

Following the idea of least squares method, we can find coefficients a_0 and a_1 by solving the following system of equations:

$$S_0 a_0 + S_1 a_1 = V_0$$

$$\hspace{6cm} (5.3)$$

$$S_1 a_0 + S_2 a_1 = V_1$$

where

$$S_0 = m+1, \quad S_1 = \sum_{i=0}^{m} x_i, \quad S_2 = \sum_{i=0}^{m} x_i^2,$$

$$V_0 = \sum_{i=0}^{m} y_i, \quad V_1 = \sum_{i=0}^{m} x_i y_i.$$

The function

$$D(a_0, a_1) = \sum_{i=0}^{m} (y_i - a_0 - a_1 x_i)^2$$

attains its minimum at the solution $a_0^*, \; a_1^*$ of the system of equations (5.3).

Example 5.2 *Find the line of regression for the following data:*

−1	0	1	2	3	4
1	2	−1	0	1	0

Estimate the of approximation.

Solution. We have

$$m = 5, \quad n = 1, \quad S_0 = 6, \quad S_1 = 9, \quad S_2 = 31, \quad V_0 = 3, \quad V_1 = 1.$$

The coefficients a_0^* and a_1^* of the line of regression

$$P_1(x) = a_0^* + a_1^* x$$

are determined by the following system of two equations:

$$6a_0 + 9a_1 = 3$$
$$9a_0 + 31a_1 = 1. \tag{5.4}$$

Solving this system of equations, we find the coefficients

$$a_0^* = \frac{4}{5}, \quad a_1^* = -\frac{1}{5}$$

and the equation of the line of regression

$$P_1(x) = \frac{4}{5} - \frac{1}{5}x.$$

In the example. the root mean square error is

$$E = \sqrt{\frac{D(0.8, -0.2)}{5 + 1}} = \sqrt{\frac{4.64}{6}} = 0.879$$

To find the line of regression through data points, we execute the following commands:

```
data={-1,1},{0,2},{1,-1},{2,0},{3,1},{4,0}};
Fit[data,{1,x},x];
```

5.3 Functions Determined by Experimental Data

A relationship between two physical phenomena can be found by the least squares method using experimental data. Namely, let us assume that y depends on x in the following way:

$$y = F(x; a_0, a_1, a_2, \ldots, a_n),$$

where F is a given function of the variable x and $(n + 1)$ parameters $a_0, a_1, a_2, \ldots, a_n$. Within this $(n + 1)$ parameter family of functions, we shall find one which is the best fitted to the experimental data (*cf.* [2,3,16,17]):

x_0	x_1	x_2	x_3	$\cdots$	$\cdots$	$\cdots$	x_m
y_0	y_1	y_2	y_3	$\cdots$	$\cdots$	$\cdots$	y_m

The parameters $a_0, a_1, a_2, \ldots, a_n$ should satisfy the following conditions:

$$y_0 = F(x_0, a_0, a_1, \ldots, a_n),$$

$$y_1 = F(x_1, a_0, a_1, \ldots, a_n)$$

$$\cdots\cdots\cdots\cdots\cdots\cdots\cdots$$

$$y_m = F(x_m, a_0, a_1, \ldots, a_n)$$

(5.5)

The system of nonlinear equations (5.5) may not have a solution when $m > n$. However, we can determine such coefficients $a_0^*, a_1^*, \ldots, a_n^*$ for which the function

$$D(a_0, a_1, \ldots, a_n) = \sum_{i=0}^{m} [y_i - F(x_i, a_0, a_1, \ldots, a_n)]^2,$$

attains its minimum in the whole real space R^{n+1}.
If $F(x, a_0, a_1, \ldots, a_n)$ is continuously differentiable with respect to parameters $a_0, a_1, \ldots, a_n$, then the function

$$D(a_0, a_1, \ldots, a_n)$$

attains its minimum at $a_0^*, a_1^*, \ldots, a_n^*$, for which

$$\frac{\partial D(a_0^*, a_1^*, \ldots, a_n^*)}{\partial a_k} = 0, \quad k = 0, 1, \ldots, n.$$

Thus, we have to solve the following nonlinear algebraic system of equations:

$$\sum_{i=0}^{m}[y_i - F(x_i, a_0, a_1, \ldots, a_n)]\frac{\partial F(x_i, a_0, a_1, \ldots, a_n)}{\partial a_k} = 0,$$

$$(5.6)$$

$$k = 0, 1, \ldots, n.$$

Example 5.3 *Let*

$$F(x, a_0, a_1) = a_0\, e^{-a_1\, x}.$$

From an experiment the following data have been determined:

x	0	1	2
$y = F$	3	1	0.5

Find in the family $F(a_0, a_1)$ the best fitted function to the given data in the table.

In this example $m = 2, \ \ n = 1$ and $F(x, a_0, a_1) = a_0\, e^{-a_1\, x}$. Then, the system of equations (5.6) takes the following form:

$$\sum_{i=0}^{2}[y_i - a_0\, e^{-a_1\, x_i}]e^{-a_1\, x_i} = 0$$

$$(5.7)$$

$$\sum_{i=0}^{2}[-y_i + a_0\, e^{-a_1\, x_i}]a_0\, x_i\, e^{-a_1\, x_i} = 0$$

Hence, we get

$$(3 - a_0) + (1 - a_0\, e^{-a_1})\, e^{-a_1} + (0.5 - a_0\, e^{-2a_1})\, e^{-2a_1} = 0$$

$$(a_0\, e^{-a_1} - 1)\, a_0\, e^{-a_1} + 2\, a_0\, (a_0\, e^{-2a_1} - .5)\, e^{-2a_1} = 0$$

$$(5.8)$$

Let $z = e^{-a_1}$ and $a = a_0$, then, we may rewrite the system of equations (5.8) as follows:

$$a = 3 + z + (0.5 - a)z^2 - az^4$$

$$z = \frac{1 - 2az^3}{a - 1} \tag{5.9}$$

Applying the iterative method

$$a^{(0)} = 0, \quad z^{(0)} = 0,$$

$$a^{(s+1)} = 3 + z^{(s)} + (0.5 - a^{(s)})[z^{(s)}]^2 - a^{(s)}[z^{(s)}]^4,$$

$$z^{(s+1)} = \frac{1 - 2\,a^{(s+1)}[z^{(s)}]^3}{a^{(s+1)} - 1}, \quad s = 0, 1, \ldots;$$

we arrive at the solution $a = 2.985652$, $z = 0.36152142$. Hence

$$a_0^* = 2.985652, \quad a_1^* = -ln\,z = 1.017434.$$

Then, the best fitted function to data is

$$F(x, a_0^*, a_1^*) = 2.985652\,e^{-1.017434\,x}$$

Now, we shall find a function within the family $F(x, a_0, a_1, \ldots, a_n)$ by the least squares method.

Let $a_0^{(0)}, a_1^{(0)}, \ldots, a_n^{(0)}$ be a solution of $(n+1)$ first equations $m > n$. If $F(x, a_0, a_1, \ldots, a_n)$ is a continuously differentiable function with respect to $a_0, a_1, \ldots, a_n$, then

$$F(x_k, a_0, a_1, \ldots, a_n) \approx F(x_k, a_0^{(0)}, a_1^{(0)}, \ldots a_n^{(0)})$$

$$+ \sum_{i=0}^{m} \frac{\partial F(x_k, a_0^{(0)}, a_1^{(0)}, \ldots, a_n^{(0)})}{\partial a_i} \alpha_i$$

for $k = 0, 1, \ldots, n$. Denoting by

$$b_{ki} = \frac{\partial F(x_k, a_0^{(0)}, a_1^{(0)}, \ldots, a_n^{(0)})}{\partial a_i},$$

and
$$l_k = y_k - F(x_k, a_0^{(0)}, a_1^{(0)}, \ldots, a_n^{(0)}),$$
we can rewrite the system of equations (5.5) in the following form:

$$b_{00}\alpha_0 + b_{01}\alpha_1 + \cdots + b_{0n}\alpha_n = l_0$$

$$b_{10}\alpha_0 + b_{11}\alpha_1 + \cdots + b_{1n}\alpha_n = l_1$$

$$\ldots\ldots\ldots\ldots\ldots\ldots\ldots\ldots \tag{5.10}$$

$$b_{m0}\alpha_0 + b_{m1}\alpha_1 + \cdots + b_{mn}\alpha_n = l_m$$

In general, the system of equations (5.10) does not have a solution when $m > n$. However, it is possible to find such

$$\alpha_0^*, \alpha_1^*, \ldots, \alpha_n^*$$

for which the function

$$D(\alpha_0, \alpha_1, \ldots, \alpha_n) = \sum_{k=0}^{m} p_k [l_k - \sum_{i=0}^{m} b_{ki}\alpha_i]^2$$

attains its minimum at $\alpha_0^*, \alpha_1^*, \ldots, \alpha_n^*$, where

$$p_k = \frac{\mu_k}{\mu}, \quad \mu = \sqrt{\frac{\mu_0^2 + \mu_1^2 + \cdots + \mu_m^2}{m+1}},$$

and μ_k is an average error related with evaluation of y_k.
The function $D(\alpha_0, \alpha_1, \ldots, \alpha_n)$ has minimum at $\alpha_0^*, \alpha_1^*, \ldots, \alpha_n^*$ if

$$\frac{\partial D(\alpha_0^*, \alpha_1^*, \ldots, \alpha_n^*)}{\partial \alpha_k} = 0, \quad k = 0, 1, \ldots, n.$$

We can find $\alpha_0^*, \alpha_1^*, \ldots, \alpha_n^*$ solving the following system of linear equations:

$$\sum_{i=0}^{m} (p_i \sum_{k=0}^{m} b_{ki} b_{kj})\alpha_i = \sum_{i=0}^{m} p_i l_i b_{kj} \tag{5.11}$$

$j = 0, 1, \ldots, n.$
Now, we set

$$a_i^* = a_i^0 + \alpha_i^*, \quad i = 0, 1, \ldots, n,$$

as approximate values of the parameters $a_0^*, a_1^*, \ldots, a_n^*$. Then, the best fitted function to the data table is

$$y = F(x, a_0^*, a_1^*, \ldots, a_n^*).$$

Example 5.4 *Let us apply the above method to find*

$$F(x, a_0, a_1) = a_0 \, e^{-a_1 x}$$

for the data as in example 1.

x	0	1	2
$y = F$	3	1	0.5

Obviously, we have

$$m = 2, \quad n = 1, \quad F(a_0, a_1) = a_0 \, e^{-a_1 x}$$

and

$$y_0 = a_0 e^{-a_1 x_0}, \qquad\qquad a_0 = 3$$

$$y_1 = a_0 e^{-a_1 x_1}, \qquad a_0 e^{-a_1} = 1 \qquad (5.12)$$

$$y_2 = a_0 e^{-a_1 x_2}, \quad a_0 e{-2a_1} = 0.5$$

We can set the initial values

$$a_0^{(0)} = 3, \qquad a_1^{(0)} = ln\, 3 = 1.098612.$$

as the solution of two first equations in (5.12).
Then, we have

$$b_{00} = e^{-a_1^{(0)} x_0} = 1, \quad b_{01} = -x_0 a_0^{(0)} e^{-a_1^{(0)} x_0} = 0,$$

$$l_0 = y_0 - a_0^{(0)} e^{-a_1^{(0)} x_0} = 0,$$

$$b_{10} = e^{-a_1^{(0)} x_1} = \frac{1}{3}, \quad b_{11} = -x_1 a_0 e^{-a_1 x_1} = -1,$$

$$l_1 = y_1 - a_0 e^{-a_1 x_1} = 0,$$

$$b_{20} = e^{-a_1 x_2} = \frac{1}{9}, \quad b_{21} = -x_2 a_0 e^{-a_1 x_2} = -\frac{2}{3},$$

$$l_2 = y_2 - a_0 e^{-a_1 x_2} = \frac{1}{6}.$$

Thus, the system of equations (5.10) takes the following form:

$$\alpha_0 = 0$$

$$\tfrac{1}{3}\alpha_0 - \alpha_1 = 0 \qquad\qquad (5.13)$$

$$\tfrac{1}{9}\alpha_0 - \tfrac{2}{3}\alpha_1 = \tfrac{1}{6}$$

So that, the function

$$D(\alpha_0, \alpha_1) = \alpha_0^2 + (\tfrac{1}{3}\alpha_0 - \alpha_1)^2 + (\tfrac{1}{6} - \tfrac{1}{9}\alpha_0 + \tfrac{2}{3}\alpha_1)^2$$

attains its minimum if

$$\frac{\partial D}{\partial a_0} = \frac{182}{81}\alpha_0 - \frac{22}{27}\alpha_1 - \frac{1}{27} = 0,$$

$$\frac{\partial D}{\partial a_1} = -\frac{22}{27}\alpha_0 + \frac{26}{9}\alpha_1 + \frac{2}{9} = 0.$$

Hence

$$182\alpha_0 - 198\alpha_1 = 3,$$

$$-22\alpha_0 + 78\alpha_1 = -6.$$

and

$$\alpha_0^* = -0.09695122, \qquad \alpha_1^* = -0.1042683.$$

Finally, we determine the coefficients

$$a_0^* = a_0^{(0)} + \alpha_0^* = 2.9030488, \qquad a_1^* = a_1^{(0)} + \alpha_1^* = 0.9943437,$$

and the function

$$F(x, a_0^*, a_1^*) = 2.9030488e^{-0.9943437x}.$$

Example 5.5 *The function $f(x) = e^{-x}(2 + \sin \pi x)$ is tabulated below.*

x	1	1.2	1.4	1.6	1.8	2
$y = F$	0.735759	0.425351	0.258666	0.211778	0.233438	0.270671

Use `Mathematica Fit` function to find the best fitted exponential function in the two parameters family

$$F(x, a_0, a_1) = e^{a_0 + a_1 x}$$

.

```
data=N[Table[{x,Exp[-x]*(2+Sin[Pi x])},{x,1,2,0.2}]];
    Exp[Fit[Log[data],{1,x},x]];
```

we obtain the best fitted exponential function

$$g(x) = e^{-0.551954-1.54912\,x} = 0.575824e^{-1.54912x}.$$

5.4   Least Squares Method in Space $L_2(a,b)$

The spaces $L_2(a,b)$ **and** $L_2^\rho(a,b)$. Let us consider the space of all functions integrable with squares on interval $[a,b]$, so that

$$L_2(a,b) = \{f \; : \; \int_a^b f^2(x)dx < +\infty\}.$$

The inner product of two functions $f, g \in L_2(a,b)$ is defined as follows:

$$(f,g) = \int_a^b f(x)g(x)dx,$$

and this product generates the norm

$$||f|| = \sqrt{(f,f)} = \sqrt{\int_a^b f^2(x)dx}, \qquad f \in L_2(a,b).$$

The space $L_2(a,b)$ is naturally associated with the following space (*cf.* [2, 18, 19])

$$L_2^\rho(a,b) = \{f \; : \; \int_a^b \rho(x)f^2(x)dx < +\infty\}.$$

where weight [1] $\rho(x) \geq 0$. The inner product in $L_2^\rho(a,b)$ space

$$(f,g)_\rho = \int_a^b \rho(x)f(x)g(x)dx, \qquad f,g \in L_2^\rho(a,b),$$

and the norm

$$||f||_\rho = \sqrt{(f,f)_\rho}, \qquad f \in L_2^\rho(a,b).$$

---

[1]It is assumed that the weight $\rho(x) \neq 0$ almost everywhere

Consequently, two functions f and g are called orthogonal either in $L_2(a, b)$ space or in $L_2^\rho(a, b)$ if

$$(f, g) = 0 \quad \text{or} \quad (f, g)_\rho = 0.$$

A system of functions

$$\phi_0(x), \phi_1(x), ..., \phi_n(x), ...,$$

is orthogonal if

$$(\phi_m, \phi_n) = 0 \quad \text{for} \quad m \neq n,$$

and it is called orthonormal if

$$(\phi_m, \phi_n) = \begin{cases} 1 & m = n, \\ 0 & m \neq n, \end{cases}$$

Example 5.6 *The system of trigonometric functions*

$$1, \ \sin\frac{\pi x}{l}, \ \cos\frac{\pi x}{l}, \ \sin\frac{2\pi x}{l}, \ \cos\frac{2\pi x}{l}, \ ..., \ \sin m\frac{\pi x}{l}, \ \cos\frac{m\pi x}{l}, ...,$$

is orthogonal in interval $[-l, l]$.

Indeed, we note that

$$\int_{-l}^{l} \sin\frac{m\pi x}{l} dx = 0, \qquad \int_{-l}^{l} \cos\frac{m\pi x}{l} dx = 0$$

for $m = 1, 2, ...,$
and by the trigonometric identities

$$\sin\frac{m\pi x}{l} \cos\frac{n\pi x}{l} = \frac{1}{2}[\sin\frac{\pi(m-n)}{l}x + \sin\frac{\pi(m+n)}{l}x],$$

$$\sin\frac{m\pi x}{l} \sin\frac{n\pi x}{l} = \frac{1}{2}[\cos\frac{\pi(m-n)}{l}x - \cos\frac{\pi(m+n)}{l}x]$$

$$\cos\frac{m\pi x}{l} \cos\frac{n\pi x}{l} = \frac{1}{2}[\cos\frac{\pi(m-n)}{l}x + \cos\frac{\pi(m+n)}{l}x],$$

we find

$$\int_{-l}^{l} \sin\frac{m\pi x}{l} \sin\frac{n\pi x}{l} dx = \begin{cases} l & m = n, \\ 0 & m \neq n, \end{cases}$$

$$\int_{-l}^{l} \sin\frac{m\pi x}{l}\cos\frac{n\pi x}{l}dx = \begin{cases} l & m=n, \\ 0 & m\neq n, \end{cases}$$

$$\int_{-l}^{l} \cos\frac{m\pi x}{l}\cos\frac{n\pi x}{l}dx = \begin{cases} l & m=n, \\ 0 & m\neq n. \end{cases}$$

Hence, the system of trigonometric functions is orthogonal in $L_2(-l, l)$ space.

Clearly, the system of functions

$$\frac{1}{2l}, \quad \frac{1}{l}\sin\frac{\pi x}{l}, \quad \frac{1}{l}\cos\frac{\pi x}{l}, \quad \frac{1}{l}\sin\frac{2\pi x}{l}, \quad ..., \quad \frac{1}{l}\sin m\frac{\pi x}{l}, ...$$

is orthonormal in space $L_2(-l, l)$.

Orthogonal functions play an important role in the theory and applications of the least squares method. Therefore, we shall present some of known systems of orthogonal functions.

Orthogonal polynomials. Below, we shall list orthogonal polynomials in space $L_2^{\rho}(a, b)$.

1. (a) *Chebyshev Polynomials*

$$\overline{T}_n(x) = \frac{1}{2^{n-1}}\cos(n\ arccos\ x), \qquad n = 0, 1, ...,$$

$$[\overline{T}]_0(x) = 2, \quad [\overline{T}]_1(x) = x, \quad [\overline{T}]_2(x) = x^2 - \frac{1}{2}, \quad ...,$$

are orthogonal in space $L_2^{\rho}(-1, 1)$ with the weight

$$\rho(x) = \frac{1}{\sqrt{1-x^2}}$$

, so that

$$\int_{-1}^{1} \frac{[\overline{T}]_m(x)[\overline{T}]_n(x)}{\sqrt{1-x^2}}dx = \begin{cases} \dfrac{\pi}{2^{2m-1}}, & m=n, \\ 0, & m\neq n. \end{cases}$$

Indeed, let $x = \cos t$. Then, we have

$$\int_{-1}^{1} \frac{T_m(x)T_n(x)}{\sqrt{1-x^2}}dx = \frac{1}{2^{m+n-2}}\int_0^{\pi}\cos mt\ \cos nt\ dt$$

$$= \begin{cases} 4\pi, & m = n = 0, \\ \dfrac{\pi}{2^{2m-1}}, & m = n \neq 0, \\ 0, & m \neq n. \end{cases}$$

(b) *Hermite Polynomials* take the following forms:

$$H_n(x) = (-1)^n e^{x^2} \frac{d^n e^{-x^2}}{dx^n},$$

or after differentiation

$$H_n(x) = \sum_{k=0}^{[n/2]} \frac{(-1)^k n!}{k!(n-2k)!}(2x)^{n-2k} \qquad n = 0, 1, ...,$$

Hence

$$H_0(x) = 1, \qquad H_1(x) = 2x,$$

$$H_2(x) = 4x^2 - 2, \quad H_3(x) = 8x^3 - 12x, \ ...,$$

These polynomials are orthogonal in space $L_2^\rho(-\infty, \infty)$ with the weight $\rho(x) = e^{-x^2}$, so that

$$\int_{-\infty}^{\infty} e^{-x^2} H_m(x) H_n(x) dx = \begin{cases} 2^m m! \sqrt{\pi}, & m = n, \\ 0, & m \neq n. \end{cases}$$

Let us note that

$$\int_{-\infty}^{\infty} e^{-x^2} H_m(x) H_n(x) dx = (-1)^n \int_{-\infty}^{\infty} H_m(x) \frac{d^n e^{-x^2}}{dx^n} dx,$$

Integrating by parts, we obtain

$$(-1)^n \int_{-\infty}^{\infty} H_m(x) \frac{d^n e^{-x^2}}{dx^n} dx$$

$$= (-1)^n H_m(x) \frac{d^{n-1} e^{-x^2}}{dx^{n-1}} \Big|_{-\infty}^{\infty}$$

$$-(-1)^n \int_{-\infty}^{\infty} H'_m(x) \frac{d^{n-1} e^{-x^2}}{dx^{n-1}} dx$$

$$= (-1)^{n+1} \int_{-\infty}^{\infty} H'_m(x) \frac{d^{n-1} e^{-x^2}}{dx^{n-1}} dx$$

$$\dots\dots\dots\dots\dots\dots\dots\dots\dots\dots\dots\dots\dots$$

$$= (-1)^{2n} \int_{-\infty}^{\infty} H_m^{(n)}(x) e^{-x^2} dx.$$

Because $H_m^{(n)}(x) = 0$ for $n > m$, therefore

$$\int_{-\infty}^{\infty} e^{-x^2} H_m(x) H_n(x) dx = 0, \quad m \neq n.$$

Using the equality

$$H_m^{(n)}(x) = 2^n n! \quad \text{for} \quad m = n,$$

we compute

$$\int_{-\infty}^{\infty} e^{-x^2} H_m(2) dx = 2^n n! \int_{-\infty}^{\infty} e^{-x^2} dx = 2^n n! \sqrt{\pi}.$$

(c) *Laguerre Polynomials* are given by the following formulas:

$$LE_n(x) = \frac{e^x}{n!} \frac{d^n e^{-x} x^n}{dx^n},$$

or after differentiation

$$LE_n(x) = \sum_{k=0}^{n} \binom{n}{k} \frac{(-x)^k}{k!}, \quad n = 0, 1, ...,$$

Hence

$$LE_0(x) = 1, \quad LE_1(x) = 1-x, \quad LE_2(x) = 1-2x+\frac{1}{2}x^2; ...,$$

Laguerre polynomials are orthogonal in space $L_2^\rho(0, \infty)$ with the weight $\rho(x) = e^{-x}$, so that

$$\int_0^\infty e^{-x} LE_m(x) LE_n(x) dx = \begin{cases} 1, & m = n, \\ 0, & m \neq n. \end{cases}$$

(d) *Legendre Polynomials* are given by the following formula:

$$LG_n(x) = \frac{1}{2^n n!} \frac{d^n (x^2 - 1)^n}{dx^n}, \qquad n = 0, 1, ...,$$

$$LG_0(x) = 1, \quad LG_1(x) = x, \quad LG_2(x) = \frac{3}{2} x^2 - \frac{1}{2}, \quad ...,$$

These polynomials are orthogonal in space $L_2(-1.1)$ and

$$\int_{-1}^1 LG_m(x) LG_n(x) dx = \begin{cases} \dfrac{2}{2n + 1}, & m = n, \\ 0, & m \neq n. \end{cases}$$

Gram - Schmidt orthogonalization. Let

$$\phi_0(x), \phi_1(x), ..., \phi_n(x), ...,$$

be a system of linearly independent functions with $\phi_k \in L_2(a, b) \cup L_2^\rho(a, b)$, for $k = 0, 1, ...$ Such a system of functions can be transformed into the orthogonal one

$$\psi_0(x), \ \psi_1(x), \ ..., \ \psi_n(x), ...,$$

using the following Gram-Schmidt algorithm:

$$\psi_0 = \phi_0,$$

for $m = 1, 2, ...,$

evaluate

$$\psi_m = \phi_m - \sum_{k=0}^{m-1} \frac{(\phi_m, \psi_k)}{(\psi_k, \psi_k)} \psi_k$$

Indeed, we note that

$$(\psi_0, \psi_0) = (\phi_0, \phi_0) = ||\phi_0||^2,$$

$$(\psi_1, \psi_0) = (\phi_1, \psi_0) - \frac{(\phi_1, \psi_0)}{(\psi_0, \psi_0)}(\psi_0, \psi_0) = 0,$$

and by mathematical induction

$$(\psi_m, \psi_n) = (\phi_m, \psi_n) - \sum_{k=0}^{m-1} \frac{(\phi_m, \psi_k)}{(\psi_k, \psi_k)}(\psi_k, \psi_n).$$

Since $(\psi_k, \psi_n) = 0$ for $k \neq n$, and $k, n < m$, therefore

$$(\psi_m, \psi_n) = (\phi_m, \psi_n) - \frac{(\phi_m, \psi_n)}{(\psi_n, \psi_n)}(\psi_n, \psi_n) = 0.$$

An implementation of Gram-Schmidt algorithm in a computer may lead to a non-orthogonal system because of round-off errors accumulation. Nevertheless, one can minimized the affect of round-off errors by applying the algorithm to the newly obtained system of functions, repeatedly.

Example 5.7 *The polynomials:* $\phi_n(x) = x^n$, $n = 0, 1, ...,$ *are linearly independent for all real* x.

Applying Gram-Schmidt algorithm to these polynomials in the space $L_2(-1, 1)$, we arrive at orthogonal polynomials which are different from the Legendre polynomials listed above by a constant factor, so that

$$\psi_0(x) = \frac{1}{\sqrt{2}}, \ \psi_1(x) = \sqrt{\frac{3}{2}}x,$$

$$\psi_2(x) = \frac{3}{2}\sqrt{\frac{5}{2}}(x^2 - \frac{1}{3}), \psi_3(x) = \frac{5}{2}\sqrt{\frac{7}{2}}(x^3 - \frac{3}{5}x), \ ...,$$

We can use `Mathematica` module `GramSchmidt` (*cf.* [28]) to transform a system of linearly independent functions into a system of orthogonal functions. For example, to transform polynomials $1, x, x^2, x^3, x^4$ into Legendre polynomials, we execute the following commands:

```
Needs["LinearAlgebra'Master'"];
GramSchmidt[{1,x,x^2,x^3,x^4},InnerProduct->
(Integrate[#1 #2,{x,-1,1}]&)]//Simplify
```

The best least squares polynomial. Let us consider a function $f \in L_2^{\rho}(a, b)$ and the generalized polynomial

$$P_n(x) = a_0\phi_0(x) + a_1\phi_1(x) + a_2\phi_2(x) + \cdots + a_n\phi_n(x).$$

In order to approximate f by a polynomial of degree not greater than n, we shall minimize the function

$$D(a_0, a_1, ..., a_n) = ||f - P_n||^2 = \int_a^b \rho(x)[f(x) - P_n(x)]^2 dx.$$

The function D attains its minimum at a point a_0^*, a_1^*, ,..., a_n^* if the partial derivatives

$$\frac{\partial D(a_0^*, a_1^*, ..., a_n^*)}{\partial a_k} = 0, \quad k = 0, 1, ..., n.$$

On the other hand

$$\frac{\partial D(a_0^*, a_1^*, ..., a_n^*)}{\partial a_k} = -2\int_a^b \rho(x)[f(x) - P_n(x)]\phi_k(x) \, dx.$$

Hence, the coefficients of the best approximating polynomial must satisfy the following system of linear algebraic equations:

$$(\phi_0, \phi_0)a_0 + (\phi_1, \phi_0)a_1 + \cdots + (\phi_n, \phi_0)a_n \quad = (f, \phi_0)$$

$$(\phi_0, \phi_1)a_0 + (\phi_1, \phi_1)a_1 + \cdots + (\phi_n, \phi_1)a_n \quad = (f, \phi_1)$$

$$\cdots\cdots\cdots\cdots\cdots\cdots\cdots\cdots\cdots\cdots\cdots\cdots\cdots\cdots\cdots\cdots \quad \cdots\cdots$$

$$\cdots\cdots\cdots\cdots\cdots\cdots\cdots\cdots\cdots\cdots\cdots\cdots\cdots\cdots\cdots\cdots \quad \cdots\cdots$$

$$(\phi_0, \phi_n)a_0 + (\phi_1, \phi_n)a_1 + \cdots + (\phi_n, \phi_n)2na_n \quad = (f, \phi_n)$$

$$(5.14)$$

where

$$(\phi_k, \phi_m) = \int_a^b \rho(x)x^{k+m}dx, \qquad (f, \phi_k) = \int_a^b \rho(x)x^k f(x)\,dx.$$

The above system of equations has a unique solution, since its Gram matrix of linearly independent functions is non-singular. Thus, the solution $a_0^*, a_1^*, ..., a_n^*$ determines the best least squares polynomial approximating function f on interval $[a, b]$. In order to estimate the error of approximation the following formula can be used:

$$E(f, P_n) = \frac{1}{b-a}\sqrt{D(a_0^*, a_1^*, ..., a_n^*)}.$$

In particular, let us consider the polynomial

$$P_n(x) = a_0 + a_1 x + a_2 x^2 + \cdots + a_n x^n.$$

The coefficients of this polynomial must satisfy the system of equations

$$\frac{\partial D(a_0^*, a_1^*, ..., a_n^*)}{\partial a_k} = -2\int_a^b \rho(x)[f(x) - P_n(x)]x^k\,dx = 0.$$

This system of equations has the following explicit form:

$$s_0 a_0 + s_1 a_1 + \cdots + s_n a_n \qquad = m_0$$

$$s_1 a_0 + s_2 a_1 + \cdots + s_{n+1} a_n \qquad = m_1$$

$$\dots\dots\dots\dots\dots\dots\dots\dots\dots\dots \qquad \dots\dots \qquad\qquad (5.15)$$

$$\dots\dots\dots\dots\dots\dots\dots\dots\dots\dots \qquad \dots\dots$$

$$s_n a_0 + s_{n+1} a_1 + \cdots + s_{2n} a_n \qquad = m_n$$

where

$$s_k = \int_a^b \rho(x)x^k dx, \qquad m_k = \int_a^b \rho(x)x^k f(x)\,dx, \qquad k = 0, 1, ...,$$

Example 5.8 *Apply the least squares method either in $L_2(-1,1)$ space or in $L_2^\rho(-1,1)$ space to approximate the function:*

$$f(x) = \sqrt{1 - x^2}, \qquad -1 \le x \le 1,$$

by a polynomial $P_2(x) = a_0\phi_0(x) + a_1\phi_1(x) + a_2\phi_2(x)$, where

1. *(a) $\phi_k(x) = x^k$, $k = 0,1,2$; and $\rho(x) = 1$, are natural polynomials*

   (b) $\phi_k(x) = T_k(x)$, $k = 0,1,2$; and $\rho(x) = \dfrac{1}{\sqrt{1-x^2}}$,
   are Chebyshev's polynomials.

Solution (a). For $\phi_k(x) = x^k$, $k = 0,1,2$; the system of linear equations (5.15) takes the following form:

$$
\begin{aligned}
2a_0 &&+\ \tfrac{2}{3}a_2 &= \frac{\pi}{2}\\[4pt]
&\tfrac{2}{3}a_1 && = 0\\[4pt]
\tfrac{2}{3}a_0 &&+\ \tfrac{2}{5}a_2 &= \frac{\pi}{8}
\end{aligned}
$$

Solving the above system of equations, we find the coefficients

$$a_0^* = \frac{21\pi}{64}, \qquad a_1^* = 0, \qquad a_2^* = -\frac{15\pi}{64}$$

and the polynomial

$$P_2(x) = \frac{3\pi}{64}(7 - 5x^2), \qquad -1 \le x \le 1.$$

Solution (b). Now, we apply the least squares method in space

$$L_2^\rho(-1,1), \quad \rho(x) = \frac{1}{\sqrt{1-x^2}},$$

where $\phi_k(x) = T_k(x)$, $k = 0,1,2$; Because Chebyshev's polynomials are orthogonal in $L_2^\rho(-1,1)$, therefore (5.14) is a diagonal system of linear equations, so that the coefficients

$$a_0^* = \frac{1}{\pi}, \qquad a_1^* = 0, \qquad a_2^* = -\frac{2\pi}{3}.$$

and the polynomial

$$P_2(x) = \frac{2}{3\pi}(2 - x^2), \quad -1 \leq x \leq 1.$$

In the table, we compare the error of approximation for the best squares polynomials and the interpolating polynomial $P_2(x) = 1 - x^2$, $x_0 = -1$, $x_1 = 0$ $x_2 = 1$, when $f(x) = \sqrt{1 - x^2}$.

Table 6.1

LS-Method	$\phi_k(x)$	$P_2(x)$	$D(a_0^*, a_1^*, a_2^*)$	$E(P_2, f)$
in $L_2(-1, 1)$	x^k	$\frac{3\pi}{64}(7 - 5x^2)$	0.00325	0.0285
in $L_2^\rho(-1, 1)$	$T_k(x)$	$\frac{2}{3\pi}(2 - x^2)$	0.42485	0.3259
Interpolation	x^k	$1 - x^2$	0.04381	0.10426

5.5   Exercises

Question 5.1 *Find the line of regression for the following data:*

-2	-1	0	1	2	3
1	0	2	3	2	4

Estimate the error of approximation.

Question 5.2 *Approximate the following data:*

0	0.25	0.5	0.75	1
0	0.707	1	0.707	0

by a polynomial of second degree. Estimate the error of approximation.

Question 5.3 *Consider the following two parameter family of functions:*

$$F(x, a_0, a_1) = a_0 + a_1 \ln(1 + x), \quad x \geq 0.$$

Find in the family the best fitted function to data:

0	1	2	3
2	2.693	3.099	3.386

Question 5.4 *Apply the least squares method either in $L_2(-2, 2)$ space or in $L_2^\rho(-2, 2)$ space to approximate the function*

$$f(x) = \sqrt{4 - x^2}, \quad -2 \leq x \leq 2,$$

by a polynomial $P_2(x) = a_0 + a_1 x + a_2 x^2$, where

1. *(a) natural polynomials $\phi_k(x) = x^k$ $k = 0, 1, 2$;   in $L_2(-2, 2)$ space,*

   (b) Chebyshev polynomials $\phi_k(x) = T_k(\frac{1}{2}x)$, $k = 0, 1, 2$; in $L_2^\rho(-2, 2)$ space, $\rho = \dfrac{1}{\sqrt{1 - x^2}}$.

Question 5.5 *Write a* Mathematica *module to approximate a given function $f(x)$ in the interval $[a, b]$ by a cubic polynomial*

Send Orders for Reprints to reprints@benthamscience.net

Lecture Notes in Numerical Analysis with Mathematica, 2014, 157-198   **157**

CHAPTER 6

Selected Methods for Numerical Integration

Abstract

Two classes of methods for numerical integration are presented : The class of Newton-Cotes methods and the class of Gauss methods. In the class of Newton-Cotes methods, the trapezoidal rule, Simpson's rule and other higher order rules like Romberg's method have been derived . The methods are clarified by examples. Mathematica modules are designed to derive Newton Cotes methods of any accuracy and to apply them for evaluation of integrals. Gauss methods are derived in general form with n Gauss-Legendres knots and in particular with n=1,2,3,4 knots. Example illustrating the methods are presented. The exact analysis of both methods of numerical integration has been carrying out. A set of questions is enclosed in the chapter.

Keywords: Newtons- Cotes formulas, Gauss formulas.

Krystyna STYŠ & Tadeusz STYŠ

6.1   Introduction

Let us consider a function $f(x)$ in an interval $[a, b]$. If $F(x)$ is an antiderivative to $f(x)$, i.e. $F'(x) = f(x)$, $x \in [a, b]$, then by the fundamental theorem of calculus

$$I(f) = \int_a^b f(x)dx = F(b) - F(a).$$

In general, the antiderivative $F(x)$ may not be known, although it may exist. For instance, the function $f(x) = e^{-x^2}$ possesses an antiderivative. However, any antiderivative to this function cannot be expressed by elementary functions. Nevertheless, we may evaluate the integral

$$\int_a^b e^{-x^2} dx$$

numerically.

There are many techniques of numerical integration based on different principles. Below, we shall present two classes of methods, *Newton-Cotes methods* and *Gauss methods* (*cf.* [3, 11,20,27]).

6.2   Newton-Cotes Methods

Let $P_n(x)$ be an interpolating polynomial to a sufficiently regular function $f(x)$ spaned by interpolating knots

$$a \leq x_0 < x_1 < \cdots < x_n \leq b.$$

Then, by the theorem 2.2, we have

$$f(x) = P_n(x) + R_n(f), \quad a \leq x \leq b,$$

where $R_n(f)$ is the error of interpolation.

Replacing the integrand $f(x)$ by its interpolating polynomial $P_n(x)$, we obtain the formula

$$\int_a^b f(x)dx = \int_a^b P_n(x)dx + E(f),$$

where

$$E(f) = \int_a^b R_n(f)dx \qquad (6.1)$$

is the truncation error.

Canceling the truncation error $E(f)$ in the above formula, we get the following approximation of the integral:

$$\int_a^b f(x)dx \approx \int_a^b P_n(x)dx.$$

Let us write the interpolating polynomial $P_n(x)$ in the Lagrange form

$$P_n(x) = l_0(x)f(x_0) + l_1(x)f(x_1) + \cdots + l_n(x)f(x_n),$$

where

$$l_i(x) = \frac{(x - x_0)(x - x_1) \cdots (x - x_{i-1})(x - x_{i+1}) \cdots (x - x_n)}{(x_i - x_0)(x_i - x_1) \cdots (x_i - x_{i-1})(x_i - x_{i+1}) \cdots (x_i - x_n)}$$

for $i = 0, 1, \ldots, n$;

Then, we have

$$\int_a^b P_n(x)dx = \sum_{i=0}^n c_i f(x_i),$$

where the coefficients

$$c_i = \int_a^b l_i(x)dx, \quad i = 0, 1, \cdots, n; \qquad (6.2)$$

are independent of f.

In this way, we arrived at the following Newton-Cotes formulas:

$$\int_a^b f(x)dx \approx \sum_{i=0}^n c_i f(x_i),$$

where the truncation error $E(f)$ is given by (6.1).

Before, we shall derive a few Newton-Cotes algorithms, let us explain the idea of simple and composed numerical methods.

All numerical methods of integration can be classified in two groups, either *simple methods or composed methods*. This classification is based on the additive property of Riemann integral. Namely, we have

$$I(f) = \int_a^b f(x)dx = \sum_{i=0}^{n-1} \int_{x_i}^{x_{i+1}} f(x)dx = \sum_{i=0}^{n-1} I_i(f).$$

Then, a numerical method is called *simple method*, if it produces approximate values of the partial results

$$I_i(f) = \int_{x_i}^{x_{i+1}} f(x)dx, \quad i = 0, 1, \ldots, n-1,$$

and a numerical method is called *composed method*, if it produces an approximate value to the integral $I(f)$ as the sum of all approximate values to individuals $I_i(f)$, $i = 0, 1, \ldots, n-1$.

6.3   Trapezian Rule.

Let us consider the linear interpolating polynomial $P_1(x)$ to the function $f(x)$ in the subinterval $[x_i, x_{i+1}]$. Then, we have

$$P_1(x) = \frac{x - x_{i+1}}{x_i - x_{i+1}} f(x_i) + \frac{x - x_i}{x_{i+1} - x_i} f(x_{i+1}).$$

and

$$f(x) = P_1(x) + \frac{f''(\xi_i(x))}{2!}(x - x_i)(x - x_{i+1})$$

for certain $\xi_i(x) \in (x_i, x_{i+1})$, and for $f(x)$ twice continuously differentiable in interval $[a, b]$.
Hence, we get

$$\begin{aligned} I_i(f) &= \int_{x_i}^{x_{i+1}} f(x)dx = \int_{x_i}^{x_{i+1}} P_1(x)dx \\ &+ \int_{x_i}^{x_{i+1}} \frac{f''(\xi_i(x))}{2!}(x - x_i)(x - x_{i+1})dx. \end{aligned}$$

where

$$\int_{x_i}^{x_{i+1}} P_1(x)dx \ = \ \frac{f(x_i)}{x_i - x_{i+1}} \int_{x_i}^{x_{i+1}} (x - x_{i+1})dx$$

$$+ \ \frac{f(x_{i+1})}{x_{i+1} - x_i} \int_{x_i}^{x_{i+1}} (x - x_i)dx$$

$$= \ \frac{x_{i+1} - x_i}{2}[f(x_i) + f(x_{i+1})]$$

$$= \ \frac{h_i}{2}[f(x_i) + f(x_{i+1})],$$

for $h_i = x_{i+1} - x_i$, $i = 0, 1, \ldots, n - 1$.

We illustrate the method on following figure (**Fig. 6.1**)

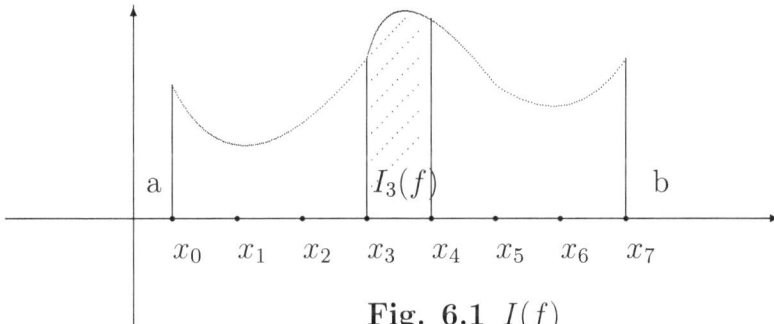

Fig. 6.1 $I(f)$

In order to estimate the error of the simple trapezian rule

$$E_i(f) = \int_{x_i}^{x_{i+1}} \frac{f''(x_i(x))}{2}(x - x_i)(x - x_{i+1})dx,$$

we note that the quadratic function $(x - x_i)(x - x_{i+1})$ does not change its sign in the interval $[x_i, x_{i+1})$. Therefore, by the mean value theorem

$$\int_{x_i}^{x_{i+1}} \frac{f''(\xi_i(x))}{2}(x-x_i)(x-x_{i+1})dx$$

$$= \frac{f''(\eta_i)}{2}\int_{x_i}^{x_{i+1}}(x-x_i)(x-x_{i+1})dx$$

$$= -\frac{f''(\eta_i)}{12}(x_{i+1}-x_i)^3.$$

for certain $\eta_i \in (x_i, x_{i+1})$.

Hence, the error of the simple trapezian rule is:

$$E_i(f) = -\frac{f''(\eta_i)}{12}(x_{i+1}-x_i)^3 = -\frac{f''(\eta_i)}{12}h_i^3.$$

for $i = 0, 1, \ldots, n-1$.

We shall consider *the composed trapezian rule* for a uniform partition of the interval $[a, b]$, *i.e.* for $x_{i+1} - x_i = h$, $i = 0, 1, \ldots, n-1$. Applying the simple trapezian rule to each of the subinterval $[x_i, x_{i+1}]$, we obtain

$$\int_a^b f(x)dx = \sum_{i=0}^{n-1}\int_{x_i}^{x_{i+1}} f(x)dx$$

$$= \frac{h}{2}\sum_{i=0}^{n-1}[f(x_i) + f(x_{i+1})] - \frac{h^3}{12}\sum_{i=0}^{n-1} f''(\eta_i).$$

By the intermediate value theorem, there exists $\eta \in (a, b)$ such that

$$\sum_{i=0}^{n-1} f''(\eta_i) = nf''(\eta) = \frac{(b-a)}{h}f''(\eta),$$

so that

$$\int_a^b f(x)dx = \frac{h}{2}[f(x_0) + \cdots + 2f(x_{n-1}) + f(x_n)] + E_T(f, h).$$

$$(6.3)$$

In this way, we arrived at the composed trapezian rule

$$T_h(f) = \frac{h}{2}[f(x_0) + 2f(x_1) + 2f(x_2) + \cdots + 2f(x_{n-1}) + f(x_n)]$$

where the truncation error

$$E_T(f, h) = -\frac{h^2}{12}(b-a)f''(\eta),$$

for certain $\eta \in (a, b)$, so that

$$I(f) = T_h(f) + E_T(f, h).$$

The truncation error of the trapezian rule satisfies the following inequality:

$$\mid E_T(f, h) \mid \leq \frac{M^{(2)}}{12}(b-a)h^2,$$

where

$$M^{(2)} = \sup_{a \leq x \leq b} \mid f''(x) \mid .$$

Example 6.1 *Evaluate the integral*

$$\int_0^2 ln(1+x)dx$$

by trapezian rule with the accuracy $\epsilon = 0.05$.

Solution. In order to get accuracy $\epsilon = 0.05$, we shall estimate the step-size h, so that, we choose the greatest $h = \dfrac{b-a}{n}$ for which the following inequality holds:

$$E_T(f, h) \leq \frac{h^2}{12}(b-a)M^{(2)} \leq \epsilon.$$

Because

$$f(x) = ln(1+x), \quad f'(x) = \frac{1}{1+x}, \quad f''(x) = -\frac{1}{(1+x)^2},$$

we have

$$M^{(2)} = \max_{0 \leq x \leq 2} \frac{1}{(1+x)^2} = 1.$$

So, the inequality

$$E_T(f, h) \leq \frac{h^2}{12} 2 < 0.05$$

holds for $h = 0.5$ and $n = 4$.
The approximate value of the integral is:

$$
\begin{aligned}
T(f) &= 0.25[f(x_0) + 2f(x_1) + 2f(x_2) + 2f(x_3) + f(x_4)] \\
&= 0.25[ln(1) + 2ln(1.5) + 2ln(2) + 2ln(2.5) + ln(3)] \\
&= 1.282105.
\end{aligned}
$$

Further error analysis. We shall give more specific analysis of the errors of trapezian method. Let us note that

$$
\begin{aligned}
f(x) &= f(y_i) + (x - y_i)f'(y_i) + \frac{(x - y_i)^2}{2} f''(y_i) \\
&+ \frac{(x - y_i)^3}{6} f'''(y_i) + \frac{(x - y_i)^4}{24} f^{(4)}(\xi_i),
\end{aligned}
$$

for certain $\xi_i \in (x_i, x_{i+1})$, and for $y_i = \frac{1}{2}(x_i + x_{i+1})$, $h = x_{i+1} - x_i$, $i = 0, 1, \ldots, n - 1$;
Hence

$$f(x_i) = f(y_i) - \frac{1}{2}hf'(y_i) + \frac{1}{8}h^2 f''(y_i) - \frac{1}{48}h^3 f'''(y_i) + \frac{1}{384}h^4 f^{(4)}(\xi_i)$$

and

$$f(x_{i+1}) = f(y_i) + \frac{1}{2}hf'(x_i) + \frac{1}{8}h^2 f''(y_i) + \frac{1}{48}h^3 f'''(y_i) + \frac{1}{384}h^4 f^{(4)}(\eta_i).$$

Therefore

$$\frac{1}{2}[f(x_i) + f(x_{i+1})] = f(y_i) + \frac{h^2}{8} f''(y_i) + \frac{1}{384} f^{(4)}(\eta_i).$$

Clearly, we have

$$\int_{x_i}^{x_{i+1}} f(x)dx = f(y_i) \int_{x_i}^{x_{i+1}} dx + f'(y_i) \int_{x_i}^{x_{i+1}} (x - y_i)dx$$

$$+ \frac{f''(y_i)}{2} \int_{x_i}^{x_{i+1}} (x - y_i)^2 dx$$

$$+ \frac{f'''(y_i)}{6} \int_{x_i}^{x_{i+1}} (x - y_i)^3 dx$$

$$+ \frac{f^{(4)}(\xi_i)}{24} \int_{x_i}^{x_{i+1}} (x - y_i)^4 dx$$

$$= hf(y_i) + \frac{h^3}{24} f''(y_i) + \frac{h^5}{1920} f^{(4)}(\xi_i)$$

Substituting to the above formula

$$f(y_i) = \frac{1}{2}[f(x_i) + f(x_{i+1})] - \frac{h^2}{8} f''(y_i) - \frac{h^4}{384} f^{(4)}(\eta_i),$$

we obtain the simple trapezian rule

$$\int_{x_i}^{x_{i+1}} f(x)dx = \frac{h}{2}[f(x_i) + f(x_{i+1})] - \frac{h^3}{12} f''(y_i) - \frac{h^5}{480} f^{(4)}(\eta_i).$$

where the truncation error

$$E_i(f, h) = -\frac{h^3}{12} f''(y_i) - \frac{h^5}{480} f^{(4)}(\eta_i).$$

Hence, we obtain the composed rule

$$\int_a^b f(x)dx = \frac{h}{2}[f(x_0) + 2f(x_1) + \cdots + 2f(x_{n-1}) + f(x_n)]$$

$$+ E_T(f, h),$$

(6.4)

which has the truncation error

$$E_T(f, h) = \sum_{i=0}^{n-1} E_i(f, h)$$

$$= -\frac{h^3}{12}[f''(y_0) + f''(y_1) + \cdots + f''(y_{n-1}]$$

$$- \frac{h^5}{480}[f^{(4)}(\eta_0) + f^{(4)}(\eta_1) + \cdots + f^{(4)}(\eta_{n-1})].$$

By the intermediate value theorem

$$f''(y_0) + f''(y_1) + \cdots + f''(y_{n-1}) = nf''(\xi) = \frac{b-a}{h}f''(\xi)$$

and

$$f^{(4)}(\eta_0) + f^{(4)}(\eta_1) + \cdots + f^{(4)}(\eta_{n-1}) = nf^{(4)}(\eta) = \frac{b-a}{h}f^{(4)}(\eta).$$

Finally, we determine the truncation error of the trapezian rule as

$$E_T(f,h) = -\frac{h^2}{12}(b-a)f''(\xi) - \frac{h^4}{480}(b-a)f^{(4)}(\eta).$$

6.4   Simpson Rule.

In order to derive Simpson rule, we consider Hermite interpolating polynomial $H_3(x)$ for the function $f(x)$ spaned by the three knots: x_{i-1}, x_i and x_{i+1} which satisfies the following conditions:

$$H_3(x_j) = f(x_j), \quad \text{for} \quad j = i-1, i, i+1,$$

and

$$H_3'(x_j) = f'(x_j), \quad \text{for} \quad j = i.$$

It is easy to check that

$$H_3(x) = L_2(x) + (x - x_{i-1})(x - x_i)(x - x_{i+1})H_1(x),$$

where Lagrange interpolating polynomial

$$
\begin{aligned}
L_2(x) \quad &= \frac{(x - x_i)(x - x_{i+1})}{(x_{i-1} - x_i)(x_{i-1} - x_{i+1})}f(x_{i-1}) \\
&+ \frac{(x - x_{i-1})(x - x_{i+1})}{(x_i - x_{i-1})(x_i - x_{i+1})}f(x_i) \\
&+ \frac{(x - x_{i-1})(x - x_i)}{(x_{i+1} - x_{i-1})(x_{i+1} - x_i)}f(x_{i+1}),
\end{aligned}
$$

and the constant polynomial

$$H_1(x) = -\frac{1}{h^2}f'(x_i) + \frac{1}{2h^3}[f(x_{i+1}) - f(x_{i-1})].$$

Let $f(x)$ be a function four times continuously differentiable in the interval $[a, b]$. By theorem 2.3, we have

$$f(x) = H_3(x) + \frac{f^{(4)}(\xi_i)}{4!}(x - x_{i-1})(x - x_i)^2(x - x_{i+1}),$$

for certain $\xi_i \in (x_{i-1}, x_{i+1})$.
Hence, we get

$$\int_{x_{i-1}}^{x_{i+1}} f(x)dx = \int_{x_{i-1}}^{x_{i+1}} H_3(x)dx$$

$$+ \frac{1}{4!}\int_{x_{i-1}}^{x_{i+1}} f^{(4)}(\xi_i(x))(x - x_{i-1})(x - x_i)^2(x - x_{i+1})dx.$$

Let us note that

$$\int_{x_{i-1}}^{x_{i+1}} (x - x_{i-1})(x - x_i)(x - x_{i+1})dx = 0.$$

So that

$$\int_{x_{i-1}}^{x_{i+1}} H_3(x)dx = \int_{x_{i-1}}^{x_{i+1}} L_2(x)dx = \frac{h}{3}[f(x_{i-1}) + 4f(x_i) + f(x_{i+1})].$$

By the mean value theorem for integrals, we get

$$\frac{1}{4!}\int_{x_{i-1}}^{x_{i+1}} f^{(4)}(\xi_i(x))(x - x_{i-1})(x - x_i)^2(x - x_{i+1})dx =$$

$$= \frac{f^{(4)}(\eta_i)}{4!}\int_{x_{i-1}}^{x_{i+1}} (x - x_{i-1})(x - x_i)^2(x - x_{i+1})dx = -\frac{f^{(4)}(\eta_i)}{90}h^5.$$

In this way, we have arrived at the simple Simpson rule

$$\int_{x_{i-1}}^{x_{i+1}} f(x)dx = \frac{h}{3}[f(x_{i-1}) + 4f(x_i) + f(x_{i+1})]]$$
$$- \frac{f^{(4)}(\eta_i)}{90}h^5, \tag{6.5}$$

where the error of the method is:

$$E_S(f,h) = -\frac{f^{(4)}(\eta_i)}{90}h^5.$$

for certain $\eta_i \in (x_{i-1}, x_{i+1})$.

We illustrate the method on followung figure (**Fig. 6.2**)

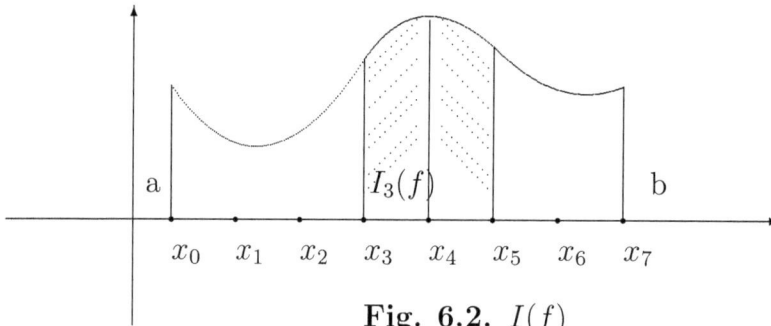

Fig. 6.2. $I(f)$

In order to derive the composed Simpson rule, we consider the uniform portion of the interval $[a,b]$ by the following points:

$$x_i = a + ih, \quad i = 0, 1, \ldots, 2n, \quad h = \frac{b-a}{2n}.$$

Applying simple Simpson rule to each term of the following sum:

$$\int_a^b f(x)dx = \int_{x_0}^{x_2} f(x)dx + \int_{x_2}^{x_4} f(x)dx + \cdots + \int_{x_{2n-2}}^{x_{2n}} f(x)dx,$$

we obtain composed Simpson rule

$$\int_a^b f(x)dx = S_h(f) + E_S(f,h),$$

where

$$S_h(f) = \frac{h}{3}[f(x_0) + 4f(x_1) + 2f(x_2)$$

$$+4f(x_3) + \cdots + 2f(x_{2n-2}) + 4f(x_{2n-1}) + f(x_{2n})],$$

and the error

$$E_S(f, h) = -\frac{h^5}{90}[f^{(4)}(\eta_1) + f^{(4)}(\eta_2) + \cdots + f^{(4)}(\eta_n)].$$

Hence, by the intermediate value theorem, there exists $\eta \in (a, b)$ such that

$$f^{(4)}(\eta_1) + f^{(4)}(\eta_2) + \cdots + f^{(4)}(\eta_n) = nf^{(4)}(\eta) = \frac{b-a}{2h}f^{(4)}(\eta).$$

Therefore, the error of the composed Simpson rule is:

$$E_S(f, h) = -\frac{h^5}{90}nf^{(4)}(\eta) = -\frac{h^4}{180}(b-a)f^{(4)}(\eta),$$

for certain $\eta \in (a, b)$.

This error satisfies the following inequality:

$$| E_S(f, h) | \leq \frac{h^4}{180}(b - a)M^{(4)}. \qquad (6.6)$$

where

$$M^{(4)} = \max_{a \leq x \leq b} | f^{(4)}(x) | .$$

Example 6.2 *Evaluate the integral*

$$I(f) = \int_0^2 ln(1 + x)dx$$

by Simpson rule using step-size $h = 0.5$. Estimate the truncation error $E_S(f, h)$.

Solution. We note that $2n = \dfrac{b-a}{h} = 4$ and

$$
\begin{aligned}
S_h(f) &= \frac{h}{3}[f(x_0) + 4f(x_1) + 2f(x_2) + 4f(x_3) + f(x_4)] \\
&= \frac{0.5}{3}[ln(1) + 4ln(1.5) + 2ln(2) + 4ln(2.5) + ln(3)] \\
&= 1.295322
\end{aligned}
$$

The exact value of $I(f) = 1.295837$, so that the error $E_S(f, h) = I(f) - S(f) = 1.295837 - 1.295322 = 0.000515$.
Also, we note that

$$f^{(4)}(x) = -\frac{6}{(1 + x)^4} \text{ and } M^{(4)} = \max_{0 \leq x \leq 2} \frac{6}{1 + x)^4} = 6.$$

Hence

$$\mid E_S(f, h) \leq \frac{h^4}{180}(b - a)M^{(4)} = \frac{0.0625}{180}2 * 6 = 0.00417.$$

6.5   Romberg Method and Recursive Formulas

Below, we shall present methods of numerical integratione based on the trapezian rule and Richartson extrapolation

Romberg Method Let us assume that the error $E_T(f, h)$ is proportional to h^2, *i.e.*

$$E_T(f, h) = C\, h^2,$$

for certain constant C.
Then, we have

$$E_T(f, 2h) = C\, 4h^2 = 4\, E_T(f, h),$$

$$I(f) = T(h, f) + E_T(f, h).$$

So that

$$I(f) = T(2h, f) + E_T(f, 2h) = T(2h, f) + 4E_T(f, h).$$

Hence, we find the truncation error

$$E(f, h) = \frac{1}{3}[T(h, f) - T(2h, f)].$$

In this way, we obtain the Romberg method

$$R_h(T, f) = \frac{1}{3}[4\, T(h, f) - T(2h, f)], \qquad (6.7)$$

for which

$$\int_a^b f(x)dx = R_h(T, f) + E_R(T, h, f),$$

where the truncation error

$$E_R(T, h, f) = E_S(f, h) = -\frac{h^4}{180}(b - a)f^{(4)}(\eta),$$

Indeed, the Romberg method has the same truncation error $E_S(f, h)$ as the Simpson rule, since

$$R_h(T, f) = \frac{1}{3}[4T_h(f) - T_{2h}(f)]$$

$$= \frac{h}{3}[f(x_0) + 4f(x_1) + 2f(x_2)$$

$$+4f(x_3) + \cdots + 4f(x_{2n-1}) + f(x_{2n})] = S_h(f).$$

Example 6.3 *Consider the following integral:*

$$\int_0^{1.2} ln(1 + x)dx.$$

(a). *Evaluate this integral by $T_h(f)$ rule with accuracy $\epsilon = 0.01$.*

(b). *Improve the result, that obtained in* **(a)**, *using the Romberg formula* **(??)**. *Estimate the error of the final result.*

Solution. We have

$$f(x) = ln(1 + x) \quad \text{and} \quad |\ f''(x)\ | = |-\frac{1}{(1 + x)^2}\ | \le 1,$$

for $x \in [0, 1, 2]$.

To determine h, we consider the inequality

$$|\ E_T(f, h)\ | \approx |-\frac{h^2}{12}(b - a)f''(\xi)\ | \le \frac{h^2}{10} \le \epsilon = 0.01,$$

for $h \le 0.31622$. So, we can choose $h = 0.3$. Then

$$n = \frac{1}{h}(b - a) = 4$$

for $b - a = 1.2$, $h = 0.3$ and $x_i = 0.3i$, $i = 0, 1, 2, 3, 4$.
By (6.4), we get

$$
\begin{aligned}
T_h(f) \quad &= \frac{h}{2}[f(x_0) + 2f(x_1) + 2f(x_2) + 2f(x_3) + f(x_4)] \\
&= 0.15[ln(1) + 2ln(1.3) + 2ln(1.6) + 2ln(1.9) + ln(1.2)] \\
&= 0.5305351.
\end{aligned}
$$

For $2h = 0.6$, we have $n = 2$ and

$$
\begin{aligned}
T_{2h}(f) \quad &= \frac{2h}{2}[f(x_0) + 2f(x_1) + f(x_2)] \\
&= 0.3[ln(1) + 2ln(1.6) + ln(2.2)] = 0.5185395.
\end{aligned}
$$

Using the Romberg formula, we compute

$$
\int_0^{1.2} ln(1 + x)dx \approx \frac{1}{3}[4T_h(f) - T_{2h}(f)] =
$$

$$
\frac{1}{3}(4 * 0.5305351 - 0.5185395) = 0.5345337.
$$

Hence, the error of the final result is:

$$
\mid E_R(f, h) \mid = \mid -\frac{h^4}{180}(b - a)f^{(4)}(\eta) \mid \leq \frac{0.3^4}{180}1.2 * 6 = 0.000324.
$$

Let us note that $\int_0^{1.2} ln(1 + x)\ dx = 0.534606$.

Recursive Formulas. Now, we consider a uniform partition of the interval $[a, b]$ into 2^n subintervals of the length $h = \dfrac{b - a}{2^n}$. The trapezian rule for such a partition is:

$$
RT(n, 0) = h \sum_{i=1}^{2^n - 1} f(a + ih) + \frac{h}{2}[f(a) + f(b)]. \qquad (6.8)
$$

From formula (6.8), we have

$$RT(0,0) \quad = \quad \frac{b-a}{2}[f(a)+f(b)],$$

$$RT(1,0) \quad = \quad hf(a+h) + \frac{h}{2}[f(a)+f(b)],$$

$$RT(2,0) \quad = \quad h[f(a+h)+f(a+2h)+f(a+3h)]$$

$$+ \quad \frac{h}{2}[f(a)+f(b)],$$

....

$$RT(n-1,0) \quad = \quad h\sum_{i=1}^{2^{n-1}-1} f(a+ih) + \frac{h}{2}[f(a)+f(b)]$$

We can compute $RT(n,0)$ using known values $f(a)$, $f(b)$ and $f(a+2h)$, $f(a+4h)$, ..., $f(a+(2^{n-1}-2)h)$, while $RT(n-1,0)$ has been evaluated. Thus, to compute $RT(n,0)$, we need to evaluate $f(x)$ at the points $a+h, a+3h, ..., a+(2^n-1)h$. These computations can be done by the following recursive formulas:

$$RT(0,0) = \frac{b-a}{2}[f(a)+f(b)], \quad h = b-a,$$

$$RT(1,0) = \frac{1}{2}RT(0,0) + hf(a+h), \quad h = \frac{b-a}{2},$$

$$RT(2,0) = \frac{1}{2}RT(1,0) + h[f(a+h)+f(a+3h)], \quad h = \frac{b-a}{4},$$

$$RT(3,0) = \frac{1}{2}RT(2,0) + h\sum_{k=1}^{4} f(a+(2k-1)h), \quad h = \frac{b-a}{8},$$

....

$$RT(n,0) = \frac{1}{2}RT(n-1,0) + h\sum_{k=1}^{2^{n-1}} f(a+(2k-1)h),$$

in general for $\quad h = \dfrac{b-a}{2^n}.$

Assuming that the error of the trapezian method has the following series expansion

$$E_T(f, h) = a_2 h^2 + a_4 h^4 + a_6 h^6 + \cdots,$$

we can accept that

$$E_T(f, h) \approx a_2 h^2.$$

Then, the Romberg metod gives the following $0(h^4)$ accurate result:

$$R_h(T, f) = RT(n, 1) = RT(n, 0) + \frac{1}{4 - 1}[RT(n, 0) - RT(n-1, 0)],$$

We shall obtain much more accurate result using the recursive formula

$$RT(n, m) = RT(n, m-1) + \frac{1}{4^m - 1}[RT(n, m-1) - RT(n-1, m-1)],$$

for $m, n \geq 1$. Using this recursive formula, we can build the following table of the Romberg's results:

$RT(0, 0)$

$RT(1, 0)$ $RT(1, 1)$

$RT(2, 0)$ $RT(2, 1)$ $RT(2, 2)$

$RT(3, 0)$ $RT(3, 1)$ $RT(3, 2)$ $RT(3, 3)$

....

$RT(n, 0)$ $RT(n, 1)$ $RT(n, 2)$ $RT(n, 3)$ .... $RT(n, n)$

Error estimate Let us assume that the error of the trapezian rule is given by the following power series in terms of $h = \dfrac{b - a}{2^n}$

$$E_T(f, h) = \int_a^b f(x)dx - RT(n, 0) = a_2 h^2 + a_4 h^4 + a_6 h^6 + \cdots,$$

$$(6.9)$$

Then, the error of the Romberg's method is

$$
\begin{aligned}
E_{RT}^{(n,1)}(f,h) &= \int_a^b f(x)\, dx - RT(n,1) \\
&= \int_a^b f(x)\, dx - [RT(n,0) \\
&\quad + \frac{1}{3}(RT(n,0) - RT(n-1,0))] \\
&= \frac{1}{3}[4E_T(f,\frac{h}{2}) - E_T(f,h)].
\end{aligned}
$$

Hence, by the assumption (6.9), we obtain power series of the error

$$
\begin{aligned}
E_{RT}^{(n,1)}(f,h) &= \int_a^b f(x)dx - RT(n,1) \\
&= -\frac{1}{4}a_4 h^4 - \frac{5}{16}a_6 h^6 - \cdots = b_4 h^4 + b_6 h^6 + \cdots.
\end{aligned}
$$

$$(6.10)$$

Hence, we obtain the following estimate of the error

$$
|\int_a^b f(x)dx - RT(n,1)| \approx \frac{1}{4}a_4 h^4, \quad h = \frac{b-a}{2^n}.
$$

In order to estimate the error

$$
\int_a^b f(x)dx - RT(n,2),
$$

we repeat the above analysis. Then, we have

$$
\begin{aligned}
E_{RT}^{(n,2)}(f,h) &= \int_a^b f(x)\, dx - RT(n,2) \\
&= \int_a^b f(x)\, dx - [RT(n,1) \\
&\quad + \frac{1}{15}(RT(n,1) - RT(n-1,1))] \\
&= \frac{1}{15}[16 E_{RT}^{(n,1)}(f,\frac{h}{2}) - E_{RT}^{(n,1)}(f,h)].
\end{aligned}
$$

So that, the error

$$E_{RT}^{(n,2)}(f,h) = \int_a^b f(x)dx - RT(n,2)$$

$$= -\frac{1}{20}b_6h^6 - \frac{1}{16}b_8h^8 - \cdots = c_6h^6 + c_8h^8 + \cdots,$$
$$(6.11)$$

for $h = \dfrac{b-a}{2^n}$.

Using the following **Mathematica** module we can obtain the elements $RT(i,j)$, $i = 0,1,...,n$, $j = 0,1,..,i$, of Romberg's table for a given function $f(x)$ in the interval $[a,b]$.

Program 6.1 *Mathematica module that gives the Romberg's table.*

```
rombergTable[f_,a_,b_,p_,q_]:=Module[{r,s,h},
h=(b-a)/2;
s[1]=f[a+h];
s[i_]:=s[i]=Module[{h=(b-a)/2^i},
Apply[Plus,Table[f[a+k h],{k,1,2^i-1,2}]]];
r[0,0]=(b-a)*(f[a]+f[b])/2;
r[1,0]=(r[0,0]+(b-a)*f[(a+b)/2])/2;
r[i_,0]:=r[i,0]=r[i-1,0]/2+(b-a)/2^i*s[i];
r[i_,j_]:=r[i,j]=
r[i,j-1]+(r[i,j-1]-r[i-1,j-1])/(4^j-1);
r[p,q]
]
```

Example 6.4 *Using th module* `rombergTable`, *build the Romberg's table*

$RT(0,0)$

$RT(1,0) \quad RT(1,1)$

$RT(2,0) \quad RT(2,1) \quad RT(2,2)$

$RT(3,0) \quad RT(3,1) \quad RT(3,2) \quad RT(3,3)$

for the function $f(x) = \ln(1 + x)$, $0 \le x \le 2$.
Give an estimate of the error

$$\int_0^2 \ln(1 + x) - RT(2, 2).$$

To obtain the Romberg's table, we define the function

```
f[x_]:=Log[1+x]
```

and execute the following `Mathematica` instruction:

```
Table[N[rombergTable[f,0,2,p,q]],{p,0,3},{q,0,p}]
                                      //TableForm
```

From `Mathematica`, we obtain the following table:

1.09861			
1.24245	1.2904		
1.2821	1.29532	1.29565	
1.29237	1.2958	1.29583	1.29583

To obtain the error estimate , we apply formula **(??)**, in the case when $n = 2$, $m = 2$, $a = 0$, $b = 2$, $h = \dfrac{1}{2}$. Then, we have

$$\left| \int_0^2 \ln(1 + x)dx - RT(2, 2) \right| \approx c_6 h^6 = 0.015635 c_6.$$

Simpson's Extrapolating Method Assuming that the error $E_S(f, h)$ of Simpson rule is proportional to h^4, *i.e.* there is a constant C such that

$$E_S(f, h) \approx Ch^4.$$

we obtain

$$E_S(f, 2h) \approx 16Ch^4 \approx 16E_S(f, h).$$

Because

$$I(f) = S_h(f) + E_S(f, h)$$

and

$$I(f) = S_{2h}(f) + E_S(f, 2h) \approx S_{2h}(f) + 16E_S(f, h),$$

therefore

$$E_S(f, h) \approx \frac{S_h(f) - S_{2h}(f)}{15}$$

In this way, we obtain the following numerical method:

$$I(f) = S_h(f) + \frac{S_h(f) - S_{2h}(f)}{15} + E_R(S, f, h).$$

Hence

$$\int_a^b f(x)dx = R_h(S, f) + E_R(S, f, h), \qquad (6.12)$$

where

$$R_h(S, f) = \frac{16S_h(f) - S_{2h}(f)}{15}.$$

In order to show that $E_R(S, f, h) = O(h^6)$, we consider the integral

$$J(x) = \int_{x_i}^x f(t)dt, \quad x \in [x_i, x_{i+1}].$$

By Taylor formula

$$J(x) = f(x_i)(x - x_i) + \frac{f'(x_i)}{2!}(x - x_i)^2 + +\frac{f''(x_i)}{3!}(x - x_i)^3$$

$$+\frac{f'''(x_i)}{4!}(x - x_i)^4 + \frac{f^{(4)}(x_i)}{5!}(x - x_i)^5 + \frac{f^{(5)}(x_i)}{6!}(x - x_i)^6$$

$$+\frac{f^{(6)}(\xi_i)}{7!}(x - x_i)^7.$$

$$(6.13)$$

Hence, we have

$$J(x_{i+1}) = \int_{x_i}^{x_{i+1}} f(t)dt = f(x_i)h + \frac{f'(x_i)}{2!}h^2 + \frac{f''(x_i)}{3!}h^3$$

$$+\frac{f'''(x_i)}{4!}h^4 + \frac{f^{(4)}(x_i)}{5!}h^5 + \frac{f^{(5)}(x_i)}{6!}h^6 + \frac{f^{(6)}(\xi_i)}{7!}h^7$$

$$(6.14)$$

for certain $\xi_i \in (x_i, x_{i+1})$.

Also, we have

$$J(x_{i-1}) = -\int_{x_{i-1}}^{x_i} f(t)dt = -f(x_i)h + \frac{f'(x_i)}{2!}h^2 - \frac{f''(x_i)}{3!}h^3$$

$$+\frac{f'''(x_i)}{4!}h^4 - \frac{f^{(4)}(x_i)}{5!}h^5 + \frac{f^{(5)}(x_i)}{6!}h^6 - \frac{f^{(6)}(\xi_i)}{7!}h^7$$

$$(6.15)$$

for certain $\xi_i \in (x_{i-1}, x_i)$.

Combining (6.14) and (6.15), we obtain the following formula:

$$\int_{x_{i-1}}^{x_{i+1}} f(x)dx = 2hf(x_i) + \frac{h^3}{3}f''(x_i) + \frac{h^5}{60}f^{(4)}(x_i) + \frac{h^7}{2520}f^{(6)}(\zeta_i)$$

$$(6.16)$$

for certain $\zeta_i \in (x_{i-1}, x_{i+1})$.

Now, let us note that

$$f(x_{i\pm1}) = f(x_i) \pm f'(x_i)h + \frac{f''(x_i)}{2!}h^2 \pm$$

$$\pm\frac{f'''(x_i)}{3!}h^3 \pm \frac{f^{(4)}(x_i)}{4!}h^4 \pm \frac{f^{(5)}(x_i)}{5!}h^5 + \frac{f^{(6)}(\eta_i)}{6!}h^6,$$

$$(6.17)$$

for certain $\eta_i \in (x_{i-1}, x_{i+1})$.

Applying (6.17), we find

$$\frac{f(x_{i-1}) - 2f(x_i) + f(x_{i+1})}{h^2} = f''(x_i) + e_i(f, h), \quad (6.18)$$

where the truncation error

$$e_i(f, h) = \frac{f^{(4)}(x_i)}{12}h^2 + \frac{f^{(6)}(\eta_i)}{360}h^4,$$

for certain $\eta_i \in (x_{i-1}, x_{i+1})$.

By (6.16) and (6.18), we have arrived at the simple Simpson rule

$$\int_{x_{i-1}}^{x_{i+1}} f(x)dx = \frac{h}{3}[f(x_{i-1}) + 4f(x_i) + f(x_{i+1})]$$

$$- \frac{h^3}{3}e_i(f, h) + \frac{h^5}{60}f^{(4)}(x_i) + O(h^7).$$

Now, we can write the error of this rule in the following form:

$$E_S(f, h, x_i) = -\frac{h^5}{36} f^{(4)}(x_i) + \frac{h^5}{60} f^{(4)}(x_i) + O(h^7)$$

$$= -\frac{f^{(4)}(x_i)}{90} h^5 + O(h^7).$$

Then, the truncation error of the composed Simpson rule is:

$$E_S(f, h) = -\frac{h^5}{90} \sum_{i=0}^{n-1} f^{(4)}(x_{2i+1}) + O(h^6).$$

In order to determine the truncation error $E_R(S, f, h)$ of the method (6.12), we note that

$$E_S(f, 2h) = -\frac{32h^5}{90} \sum_{i=0}^{\frac{n-2}{2}} f^{(4)}(x_{4i+2}) + O(h^6)$$

and

$$E_R(S, f, h) = \frac{16E_S(f, h) - E_S(f, 2h)}{15}$$

$$= -\frac{16h^5}{1350} \sum_{i=0}^{n-1} f^{(4)}(x_{2i+1}) + \frac{32h^5}{1350} \sum_{i=0}^{\frac{n-2}{2}} f^{(4)}(x_{4i+2})$$
$$+ O(h^6)$$

$$= -\frac{16h^5}{1350} \{ [f^{(4)}(x_1) - 2f^{(4)}(x_2) + f^{(4)}(x_3)]$$

$$+ [f^{(4)}(x_5) - 2f^{(4)}(x_6) + f^{(4)}(x_7)] +$$

$$+ [f^{(4)}(x_{2n-3}) - 2f^{(4)}(x_{2n-2}) + f^{(4)}(x_{2n-1})] \}$$

$$+ O(h^6).$$

By formula (6.18)

$$f^{(4)}(x_{4i+1}) - 2f^{(4)}(x_{4i+2}) + f^{(4)}(x_{4i+3}) = f^{(6)}(x_{4i+2})h^2 + O(h^4).$$

Hence, we have

$$E_R(S, f, h) = -\frac{16h^7}{1350}\{f^{(6)}(x_2) + f^{(6)}(x_6) + \cdots + f^{(6)}(x_{2n-2})\} + O(h^6).$$

By the intermediate value theorem, there exists η such that

$$f^{(6)}(x_2) + f^{(6)}(x_6) + \cdots + f^{(6)}(x_{2n-2}) = \frac{n}{2}f^{(6)}(\eta) = \frac{b-a}{4h}f^{(6)}(\eta).$$

Therefore, the truncation error of the method is:

$$E_R(S, f, h) = -\frac{4h^6}{1350}(b-a) + O(h^6) = O(h^6).$$

Example 6.5 *Evaluate the integral*

$$\int_0^2 ln(1+x)dx$$

by the modified method (6.12) using $h = 0.5$. Estimate the truncation error $E_R(S, f, h)$.

Solution. We shall apply formula (6.12). Then, for $h = 0.5$, $b - a = 2$, we have $2n = 4$ and

$$
\begin{aligned}
S_h(f) &= \frac{h}{3}[f(x_0) + 4f(x_1) + 2f(x_2) + 4f(x_3) + f(x_4)] \\
&= 1.295322,
\end{aligned}
$$

$$S_{2h}(f) = \frac{2h}{3}[f(x_0) + 4f(x_2) + f(x_4)] = 1.2904.$$

Hence, by formula (6.12)

$$R_h(S, f) = \frac{16S_h(f) - S_{2h}(f)}{15} = 1.29565.$$

The truncation error $E_R(S, f, h) = O(h^6) \approx 0.015625$. In fact, the result $R_h(S, F) = 1.29565$ is much more accurate, since the error constant in the $O(h^6)$ asymptotic estimate of the error $E_R(S, f, h)$ is small. Indeed, we have

$$I(f) = \int_0^2 ln(1+x)dx = 3ln(3) - 2 = 1.295837$$

Hence, the absolute error

$$I(f) - R_h(S, f) = 1.295837 - 1.29565 = 0.000187.$$

6.6   Gauss Methods

Introduction. In general, numerical methods of integration have the following form:

$$\int_a^b f(x)dx = C_0^n f(x_0) + C_1^n f(x_1) + \cdots + C_n^n f(x_n) + E(f),$$
(6.19)

where the coefficients $C_0^n, C_1^n, \ldots, C_n^n$; are independent of f and the error of the method $E(f) = E(f, x_0, x_1, \ldots, x_n)$ depends on f and of a distribution of the knots $x_0, x_1, \ldots, x_n$, in the interval $[a, b]$. For example, the trapezian rule has the coefficients

$$C_i^n = \begin{cases} \dfrac{1}{2}h & \text{if } i = 0, n, \\ h & \text{if } i = 1, 2, \ldots, n-1, \end{cases}$$

and the error of the method

$$E(f, h) = -\frac{h^2}{12}(b - a)f''(\eta)$$

for certain $\eta \in (a, b)$, where $h = x_{i+1} - x_i$, $i = 0, 1, \ldots, n-1$.

Let us observe that Newton-Cotes methods, based on an interpolating polynomial $P_n(x)$ spaned by knots $x_0, x_1, \ldots, x_n$, are exact $(E(f) = 0)$, when

$$f(x) = a_0 + a_1 x + a_2 x^2 + \cdots + a_m x^m$$

is a polynomial of degree $m \le n$.

Indeed, in this case $f^{(n+1)}(\xi(x)) = 0$, if $f(x)$ is a polynomial of degree $m \le n$. Then, the truncation error of Newton-Cotes methods is equal to zero, that is

$$E(f) = \frac{1}{n!} \int_a^b f^{(n+1)}(\xi(x))(x - x_0)x - x_1) \cdots (x - x_n) = 0.$$

Gauss methods are built on the principle to be the truncation error equal to zero $(E(f) = 0)$ for any polynomial f of degree $m \leq 2n + 1$.

6.7   Gauss-Legendre Knots

The points $x_0, x_1, \ldots, x_n$ are called *Gauss knots* if the formula (6.19) of numerical integration produces exact result for any polynomial

$$f(x) = a_0 + a_1 x + a_2 x^2 + \cdots + a_m x^m$$

of degree $m \leq 2n + 1$.

We can find Gauss knots applying the following theorem (*cf.* [2,3,11]):

Theorem 6.1 *For every polynomial $f_m(x)$ of degree $m \leq 2n + 1$ the truncation error of a numerical method*

$$E(f) = 0 \qquad (6.20)$$

if and only if

$$\int_a^b \omega_{n+1}(x) g_m(x) dx = 0 \qquad (6.21)$$

for each polynomial $g_m(x)$ of degree $m \leq n$, where

$$\omega_{n+1}(x) = (x - x_0)(x - x_1) \cdots (x - x_n).$$

Proof. At first, let us assume that $E(f) = 0$ for any polynomial $f_m(x)$ of degree $m \leq 2n + 1$. In particular,

$$f_m(x) = \omega_{n+1}(x) g_k(x)$$

is a polynomial of degree $m \leq 2n + 1$ provided that $g_k(x)$ is an polynomial of degree $k \leq n$. By the assumption, $E(f_m) = 0$. Evidently $\omega_{n+1}(x_i) = 0$, for $i = 0, 1, \ldots, n$.

Then, we have

$$\int_a^b f_m(x)dx = \int_a^b \omega_{n+1}(x)g_m(x)dx$$
$$= C_0^n \omega_{n+1}(x_0)g_k(x_0) + C_1^n \omega_{n+1}(x_1)g_k(x_1) + \cdots$$
$$+ C_n^n \omega_{n+1}(x_n)g_k(x_n) = 0.$$

Now, let us assume that

$$\int_a^b \omega_{n+1}(x)g_k(x)dx = 0 \qquad (6.22)$$

for each polynomial $g_k(x)$ of degree $k \leq n$

Then, the truncation error $E(f_m) = 0$ for every polynomial $f_m(x)$ of degree $m \leq 2n + 1$. Indeed, we can write the polynomial $f_m(x)$ in the following form:

$$f_m(x) = \omega_{n+1}(x)g_k(x) + r_s(x),$$

where $g_k(x)$ and $r_s(x)$ are polynomials of degree $k, s \leq n$. Then

$$f_m(x_i) = r_s(x_i), \quad i = 0, 1, \ldots, n.$$

By assumption (6.22)

$$\int_a^b f_m(x)dx = \int_a^b \omega_{n+1}(x)g_k(x)dx + \int_a^b r_s(x)dx = \int_a^b r_s(x)dx.$$
$$(6.23)$$

Since, $r_s(x)$ is apolynomial of degree $s \leq n$, we get

$$\int_a^b r_s(x)dx = C_0^n r_s(x_0) + C_1^n r_s(x_1) + \cdots + C_n^n r_s(x_n). \quad (6.24)$$

From (6.23) and (6.24)

$$\int_a^b f_m(x)dx = C_0^n f_m(x_0) + C_1^n f_m(x_1) + \cdots + C_n^n f_m(x_n).$$

and the truncation error $E(f_m) = 0$, when $f_m(x)$ is a polynomial of degree $m \leq 2n + 1$. This ends the proof.

Applying the above theorem, we can determine Gauss knots

$x_0, x_1, \ldots, x_n$ for each integer n. Namely, let us introduce the following sequence of polynomials:

$$\phi_1(x) = \int_a^x \omega_{n+1}(t)dt$$

$$\phi_2(x) = \int_a^x \phi_1(t)dt$$

$$\phi_2(x) = \int_a^x \phi_1(t)dt \qquad (6.25)$$

$$\cdots\cdots\cdots\cdots$$

$$\cdots\cdots\cdots\cdots$$

$$\phi_{n+1}(x) = \int_a^x \phi_n(t)dt$$

Also, we have

$$\int_a^b \omega_{n+1}(x)g_k(x)dx = 0$$

for every polynomial $g_k(x)$ of degree $k \le n$, (see (6.21)). On the other hand, integrating by parts, we obtain

$$\int_a^b \omega_{n+1}(x)g_k(x)dx = [\phi_1(x)g_k(x) - \phi_2(x)g_k(x) + \cdots$$

$$+(-1)^{n+1}\phi_{n+1}(x)g_k(x)]_{x=a}^{x=b} = 0$$

for every polynomial $g_k(x)$ of degree $k \le n$.
Therefore

$$\phi_i(a) = 0 \quad \text{and} \quad \phi_i(b) = 0, \quad i = 0, 1, \ldots, n+1.$$

This means that the polynomial $\phi_{n+1}(x)$ possesses the roots $x = a$ and $x = b$ of multiplicity $n + 1$, *i.e.*

$$\phi_{n+1}(x) = K \ (x-a)^{n+1}(x-b^{n+1},$$

where K is a constant.
Following the definition (see (6.25)) of $\phi_i(x)$, $i = 0, 1, \ldots, n+ 1$, we get

$$\omega_{n+1}(x) = K \ \frac{d^{n+1}(x-a)^{n+1}(x-b)^{n+1}}{d \ x^{n+1}}.$$

Hence, for $K = \dfrac{(n+1)!}{(2n+2)!}$, we arrive at Legendre polynomial

$$\omega_{n+1}(x) = \frac{(n+1)!}{(2n+2)!} \frac{d^{n+1}(x-a)^{n+1}(x-b)^{n+1}}{d\,x^{n+1}}.$$

The Legendre polynomial $\omega_{n+1}(x)$ has all roots real and distinct in the interval $[a, b]$. This can be proved by applying Rolle theorem to the polynomials $\phi_{n+1}(x), \phi_n(x), \ldots, \phi_1(x)$ and $\omega_{n+1}(x)$. Also, let us note that the roots of Lagrange polynomial $\omega_{n+1}(x)$ are symmetrically distributed about the mid-point d$=\dfrac{a+b}{2}$ of the interval $[a, b]$. Since, for $x = y+d$,

$$\begin{aligned}
\phi_{n+1}(x) &= \frac{(n+1)!}{(2n+2)!}(x-a)^{n+1}(x-b)^{n+1} \\
&= \frac{(n+1)!}{(2n+2)!}(y+d)^{n+1}(y-d)^{n+1} \\
&= \frac{(n+1)!}{(2n+2)!}(y^2-d^2)^{n+1},
\end{aligned}$$

and

$$\omega_{n+1}(y+d) = \frac{(n+1)!}{2n+2)!}\frac{d^{n+1}}{dx^{n+1}}(y^2-d^2)^{n+1}.$$

So, Gauss-Legendre methods are associated with the knots $a \le x_0 < x_1, \cdots < x_n \le b$, which are roots of Legendre polynomial $\omega_{n+1}(x)$.

Example 6.6 *Find Gauss knots in the interval* $[-1, 1]$, *when* $n = 1$ *or* $n = 2$.

Solution. For $n = 1$

$$\omega_2(x) = \frac{2!}{4!}\frac{d^2}{dx^2}(x^2-1)^2 = x^2 - \frac{1}{3} = 0$$

Hence, Gauss knots $x_0 = -\dfrac{1}{\sqrt{3}}$ and $x_1 = \dfrac{1}{\sqrt{3}}$

For $n = 2$, we determine

$$\omega_3(x) = \frac{3!}{6!} \frac{d^3}{dx^3} (x^2 - 1)^3 = x\left(x^2 - \frac{3}{5}\right) = 0.$$

So that, the Gauss knots are: $x_0 = -\sqrt{\dfrac{3}{5}}$, $x_1 = 0$, and

$x_2 = \sqrt{\dfrac{3}{5}}.$

6.8   Gauss Formulas

In order to determine the coefficients $C_0^n, C_1^n, \ldots, C_n^n$ of a Gauss formula, let us consider the following polynomials for k=0,1,...,n:

$$l_{kn}(x) = (x-x_0)(x-x_1)\cdots(x-x_{k-1})(x-x_{k+1})\cdots(x-x_n),$$

of degree n with the roots $x_0, x_1, \ldots, x_{k-1}, x_{k+1}, \ldots x_n$ which are roots of Legendre polynomial $\omega_n(x)$.

Because $l_{kn}^2(x)$ is a polynomial of degree $2n$, Gauss formula spaned on $n+1$ knots produces exact value of the integral $\int_a^b l_{kn}^2(x)dx$, i.e. the truncation error $E(l_{kn}^2) = 0$ and

$$\int_a^b l_{kn}^2(x)dx = C_0^n l_{kn}^2(x_0) + C_1^n l_{kn}^2(x_1) + \cdots + C_n^n l_{kn}^2(x_n) = C_k^n l_{kn}^2(x_k).$$

Hence, we obtain the following formula for Gauss coefficients:

$$C_k^n = \frac{\int_a^b l_{kn}^2(x)dx}{l_{kn}^2(x_k)}, \quad k = 0, 1, \ldots, n.$$

We can express the coefficients C_k^n, $k = 0, 1, \ldots, n$; in the other terms. Namely, let us observe that

$$\int_a^b (x-x_k)l_{kn}(x)l_{kn}'(x)dx = \sum_{i=0}^n C_i^n (x_i - x_k)l_{kn}(x_i)l_{kn}'(x_i) = 0,$$

since $(x - x_k)l_{kn}(x)l'_{kn}(x)$ is a polynomial of degree $2n$.
Then, we have

$$
\begin{aligned}
C_k^n l_{kn}^2(x_k) &= \int_a^b l_{kn}^2(x)dx \\
&= (x - x_k)l_{kn}^2(x) \, |_a^b - 2\int_a^b (x - x_k)l_{kn}(x)l'_{kn}(x)dx. \\
&= (b - x_k)l_{kn}^2(b) - (a - x_k)l_{kn}^2(a).
\end{aligned}
$$

(6.26)

Also, we note that

$$
(b - x_k)l_{kn}^2(b) - (a - x_k)l_{kn}^2(a) = \frac{\omega_{n+1}^2(b)}{b - x_k} - \frac{\omega_{n+1}^2(a)}{a - x_k}. \quad (6.27)
$$

By the Leibniz formula

$$
\begin{aligned}
\omega_{n+1}(x) &= \frac{(n+1)!}{(2n+2)!} \frac{d^{n+1}(x - a)^{n+1}(x - b)^{n+1}}{dx^{n+1}} \\
&= \frac{(n+1)!}{(2n+2)!} \sum_{i=0}^{n+1} \frac{(n+1)!}{i!(n-i+1)!} \frac{d^i(x - a)^{n+1}}{dx^i} \frac{d^{n-i+1}(x - b)^{n-i+1}}{dx^{n-i+1}}.
\end{aligned}
$$

(6.28)

Hence, we find

$$
\omega_{n+1}(a) = \frac{[(n+1)!]^2}{(2n+2)!}(a - b)^{n+1}
$$

$$
\omega_{n+1}(b) = \frac{[(n+1)!]}{(2n+2)!}(b - a)^{n+1}.
$$

and

$$
\begin{aligned}
\frac{\omega_{n+1}^2(b)}{b - x_k} - \frac{\omega_{n+1}^2(a)}{a - x_k} &= \\
&= \frac{[(n+1)!]^4}{[(2n+2)!]^2}(b - x_k)^{2n+2}\left(\frac{1}{b - x_k} - \frac{1}{a - x_k}\right) \\
&= \frac{[(n+1)!]^4}{[(2n+2)!]} \frac{(b - a)^{2n+3}}{(a - x_k)(b - x_k)}.
\end{aligned}
$$

(6.29)

Combining (6.26) and (6.29), we obtain the following formula for Gauss coefficients:

$$C_k^n = \frac{[(n+1)!]^4(b-a)^{2n+3}}{[(2n+2)!]^2(x_k-a)(b-x_k)(\omega_{n+1}'(x_k))^2}, \quad k = 0, 1, \ldots, n.$$
(6.30)

One can check that Gauss coefficients are positive and satisfy the following conditions of symmetry:

$$C_k^n = C_{n-k}^n, \quad k = 0, 1, \ldots, n.$$
(6.31)

6.9   Error of Gauss Methods

Let $H_{2n+1}(x)$ be Hermite interpolating polynomial satisfying the following conditions:

$$H_{2n+1}(x_i) = f(x_i), \quad H_{2n+1}'(x_i) = f'(x_i), \quad i = 0, 1, \ldots, n.$$

If $f(x)$ is $(2n+2)$ times continuously differentiable function in the interval $[a, b]$ then by the theorem 2.3

$$f(x) = H_{2n+1}(x) + \frac{f^{(2n+2)}(\eta)}{(2n+1)!}\Omega_{2n+2}(x),$$
(6.32)

for certain $\eta \in (a, b)$, where

$$\Omega_{2n+2}(x) = (x-x_0)^2(x-x_1)^2 \cdots (x-x_n)^2.$$

and $x_0, x_1, \ldots, x_n$ are roots of the Legendre polynomial $\omega_{n+1}(x)$, (see (6.27)).

Integrating both hand sides of the formula (6.32), we obtain

$$\int_a^b f(x)dx = \int_a^b H_{2n+1}(x)dx + \frac{1}{(2n+2)!}$$
$$+ \int_a^b f^{(2n+2)}(\eta(x))\Omega_{2n+2}(x)dx.$$
(6.33)

On other hand, by Gauss formula

$$\int_a^b f(x)dx = C_0^n f(x_0) + C_1^n f(x_1) + \cdots + C_n^n f(x_n) + E(f),$$
(6.34)

where Gauss coefficients C_i^n, $\quad i = 0, 1, \ldots, n$ are given by the formula (6.30) and the error of Gauss method

$$E(f) = \frac{1}{(2n+1)!} \int_a^b f^{(2n+2)}(\eta(x))\Omega_{2n+2}(x)dx.$$

Since $\Omega_{2n+2}(x) \geq 0$, $\quad x \in [a, b]$, by mean value theorem, we get

$$\int_a^b f^{(2n+1)}(\eta(x))\Omega_{2n+2}(x)dx = f^{(2n+2)}(\xi)\int_a^b \Omega_{2n+2}(x)dx.$$

for certain $\xi \in (a, b)$.
Hence, the truncation error is

$$E(f) = \frac{f^{(2n+2)}(\xi)}{(2n+2)!} \int_a^b \Omega_{2n+2}(x)dx. \qquad (6.35)$$

We can express the error $E(f)$ in the other form using the formula of integration by parts. Namely, we have

$$\begin{aligned}
\int_a^b \Omega_{2n+2}(x)dx &= -\int_a^b \phi_1(x)\omega'_{n+1}(x)dx \\
&= \int_a^b \phi_2(x)\omega''_{n+1}(x)dx \\
&\quad \cdots\cdots\cdots\cdots\cdots \\
&= (-1)^{n+1}\int_a^b \phi_{n+1}(x)\omega_{n+1}^{(n+1)}(x)dx \\
&= (-1)^{n+1}(n+1)!\int_a^b \phi_{n+1}(x)dx. \\
&\qquad\qquad\qquad\qquad\qquad (6.36)
\end{aligned}$$

and

$$\int_a^b \phi_{n+1}(x)dx = \frac{(n+1)!}{2n+2)!} \int_a^b (x-a)^{n+1}(x-b)^{n+1}dx$$

$$= -\frac{(n+1)!(n+1)}{2n+2)!} \int_a^b \frac{(x-a)^{n+2}(x-b)^{n+1}}{n+2}dx$$

$$= (-1)^{n+1}\frac{[(n+1)!]^2}{(2n+2)!} \int_a^b \frac{(x-a)^{2n+2}}{(n+2)(n+3)\cdots(2n+2)}dx$$

$$= (-1)^{n+1}\frac{[(n+1)!]^3(b-a)^{2n+3}}{[(2n+2)!]^2(2n+3)}.$$

$$(6.37)$$

From (6.36) and (6.37), we obtain the following formula for the truncation error:

$$E(f) = \frac{[(n+1)!]^4(b-a)^{2n+3}}{[(2n+2)!]^3(2n+3)}f^{(2n+2)}(\xi), \qquad (6.38)$$

for certain $\xi \in (a,b)$.

Example 6.7 *Let $n = 0$ and $[a,b] = [-1,1]$. Then, there is only one Gauss knot $x_0 = 0$, and then Gauss formula is:*

$$\int_{-1}^1 f(x)dx = 2f(0) + E(f),$$

where the error

$$E(f) = \frac{1}{3}f''(\xi)$$

for certain $\xi \in (-1,1)$.

Example 6.8 *Let $n = 1$ and $[a,b] = [-1,1]$. Then, there are two Gauss knots $x_0 = -\frac{1}{\sqrt{3}}$, $x_1 = \frac{1}{\sqrt{3}}$ and then Gauss formula is:*

$$\int_{-1}^1 f(x)dx = f(-\frac{1}{\sqrt{3}}) + f(\frac{1}{\sqrt{3}}) + E(f),$$

where the error

$$E(f) = \frac{1}{135}f^{(4)}(\xi)$$

for certain $\xi \in (-1, 1)$.

Example 6.9 *Let $n = 2$ and $[a, b] = [-1, 1]$. Then, there are three Gauss knots $x_0 = -\sqrt{\frac{3}{5}}$, $x_1 = 0$ and $x_2 = \sqrt{\frac{3}{5}}$ and then Gauss formula is:*

$$\int_{-1}^{1} f(x)dx = \frac{5}{9}f(-\sqrt{\frac{3}{5}}) + \frac{8}{9}f(0) + \frac{5}{9}f(\sqrt{\frac{3}{5}}),$$

where the error

$$E(f) = \frac{1}{15750}f^{(6)}(\xi)$$

for certain $\xi \in (-1, 1)$.

In order to apply Gauss method to an interval $[a, b]$, at first, we should transform this interval to the interval $[-1, 1]$ using the following linear transform:

$$\int_{a}^{b} f(x)dx = \frac{b-a}{2} \int_{-1}^{1} F(y)dy,$$

where

$$y = \frac{2x - a - b}{b - a}, \quad F(y) = f(\frac{1}{2}((b-a)y + a + b)),$$

Example 6.10 *Evaluate the integral*

$$\int_{0}^{2} ln(1 + x)dx \tag{6.39}$$

by Gauss formula based on four knotes. Estimate the error of the method.

Solution. At first, let us transform the integral using the linear mapping $y = x - 1$.
Then

$$\int_{0}^{2} ln(1 + x)dx = \int_{-1}^{1} ln(2 + y)dy.$$

In order to get Gauss knotes, when $n = 3$, we shall find the roots of Legendre polynomial

$$\omega_4(y) = \frac{4!}{8!} \frac{d^4(y^2 - 1)^4}{dy^4} = y^4 - \frac{6}{7}y^2 + \frac{3}{35} = 0$$

Hence, the roots are:

$$x_0 = -\sqrt{\frac{6}{7} + \sqrt{\frac{96}{245}}} = -0.8611363...,$$

$$x_1 = -\sqrt{\frac{6}{7} - \sqrt{\frac{96}{245}}} = -0.3399811...,$$

$$x_2 = \sqrt{\frac{6}{7} - \sqrt{\frac{96}{245}}} = 0.3399811...,$$

$$x_3 = \sqrt{\frac{6}{7} + \sqrt{\frac{96}{245}}} = 0.8611363...,$$

Following the formula (6.30) and (6.38), we find Gauss coefficients

$$C_0^3 = C_3^3 = \frac{[4!]^4 \, 2^9}{[8!]^2 \, (x_0 + 1)(1 - x_0)\omega_4'^2(x_0)} = 0.3478473...,$$

$$C_1^3 = C_2^3 = \frac{[4!]^4 \, 2^9}{[8!]^2 \, (x_1 + 1)(1 - x_1)\omega_4'^2(x_1)} = 0.6521451...,$$

Then, the truncation error is:

$$E(f) = \frac{[4!]^4 \, 2^9}{[8!]^2 \, 9} f^{(8)}(\xi) = \frac{1}{3472875} f^{(8)}(\xi).$$

Now, we have

$$\int_0^2 ln(1+x)dx = \int_{-1}^1 ln(2+y)dy$$

$$= 0.3478473\ ln(2 - 0.8611363)$$

$$+ 0.6521451\ ln(2 - 0.3399811)$$

$$+ 0.6521451\ ln(2 + 0.3399811)$$

$$+ 0.348473\ ln(2 + 0.8611363)$$

$$+ E(f) = 1.295837 + E(f),$$

where the truncation error

$$E(f) = \frac{1}{3472875} f^{(8)}(\xi)$$

$$= 0.0000002879\ \frac{7!}{(2+\xi)^8} \leq 0.0000002879 * 5040$$

$$= 0.00145.$$

In fact, the result is much more accurate, since the above error estimation is pessimistic one.

6.10   Exercises

Question 6.1 *Consider the integral*

$$I = \int_0^2 e^{-x^2} dx.$$

Evaluate the integral I by

1. *the trapezian rule*

2. *Simpson rule*

for the two steps $h = 0.25$ and $h = 0.5$. In the both cases, express the truncation errors $E_T(f, h)$ and $E_S(f, h)$ by the results $T_h(f)$, $T_{2h}(f)$, $S_h(f)$ and $S_{2h}(f)$.

Question 6.2 *Let $S_h(f)$ and $S_{\frac{h}{2}}(f)$ be two Simpson approximate values of the integral*

$$I(f) = \int_a^b f(x)dx$$

evaluated for h and $\dfrac{h}{2}$ with the errors $E_S(f, h)$ and $E_S(f, \dfrac{h}{2})$, respectively.

1. *Express the errors $E_S(f, h)$ and $E_S(f, \dfrac{h}{2})$ in terms of the results $S_h(f)$ and $S_{2h}(f)$.*
2. *Find a modified Simpson rule using the values $S_h(f)$ and $S_{\frac{h}{2}}$.*
3. *Evaluate the integral*

$$\int_0^2 \sqrt{1+x}\,dx$$

   by the modified Simpson rule with the accuracy $\epsilon = 0.001$.

Question 6.3 *Derive simple and composed the rectangular rules $RC_h(f, x_i)$, $i = 0, 1, \ldots, n$;   and $RC_h(f)$ based on the piecewise constant interpolating polynomial*

$$p_n(x) = f(x_0)\theta_0(x) + f(x_1)\theta_1(x) + \cdots + f(x_{n-1})\theta_{n-1}(x),$$

where $\theta_i(x)$ is the characteristic function to the interval $[x_i, x_{i+1}]$, i.e.

$$\theta_i(x) = \begin{cases} 1 & \text{for } x_i \le x \le x_{i+1}, \\ 0 & \text{for } x < x_i \text{ or } x > x_{i+1}, \end{cases}$$

for $i = 0, 1, \ldots, n - 1$.
Determine the truncation error $E_{RC}(f, h)$ of the rectangular

rule $RC_h(f)$.

Question 6.4 *Consider the simple Mid-Point method:*

$$\int_{x_i}^{x_{i+1}} f(x)dx = h \, f(\frac{x_i + x_{i+1}}{2}) + E_M^i(f,h),$$

for $i = 0, 1, ..., n-1, \quad nh = b - a$,

and the composed Mid-Point method:

$$\int_a^b f(x)dx = h \sum_{i=0}^{n-1} f(\frac{x_i + x_{i+1}}{2}) + E_M(f,h).$$

1. (a) *Determine the errors* $E_M^i(f,h)$ *and* $E_M(f,h)$.

   (b) *Find a modification of Mid-Point method based on Richardson extrapolation to get a new method of accuracy* $O(h^4)$.

   (c) *Use Mid-Point method and the modified Mid-Point method to evaluate the integral:*

   $$\int_0^2 \frac{x}{1+x}dx,$$

   with accuracy $\epsilon = 0.001$.

Question 6.5 *Build the Romberg's table*

$RT(0,0)$

$RT(1,0) \quad RT(1,1)$

$RT(2,0) \quad RT(2,1) \quad RT(2,2)$

$RT(3,0) \quad RT(3,1) \quad RT(3,2) \quad RT(3,3)$

for the function $f(x) = \dfrac{1}{1+x}, \quad 0 \le x \le 4$.
Give an estimate of the error

$$\int_0^4 \frac{1}{1+x}dx - RT(2,2).$$

Question 6.6 *Evaluate the integral*

$$\int_0^3 e^{-2x} dx$$

by Gauss method spaned by four Gauss-Legendre knots. Estimate the truncation error.

Question 6.7 *Find a Gauss method spaned by five Gauss-Legendre knots. Use the method to evaluate the integral*

$$\int_1^3 \sqrt{x}.$$

Estimate the truncation error.

Question 6.8 *Use Gauss-Legendre integrating formula based on three points to evaluate the integral:*

1. *(a)* $\int_0^1 x e^{-x} dx.$

    (b) $\int_{-\frac{\pi}{2}}^{\frac{\pi}{2}} \cos^3(x) dx$

Estimate the truncation error.

Send Orders for Reprints to reprints@benthamscience.net

Lecture Notes in Numerical Analysis with Mathematica, 2014, 199-229   **199**

CHAPTER 7

Solving Nonlinear Equations by Iterative Methods

Abstract

In this chapter the equation $F(x) = 0$, $a \leq x \leq b$, is solved by the Fix Point Iterations, Newton's Method, Secant Method and Bisection Method. The theorems on convergence and errors estimates of the methods have been stated and proved. Also, the rates of convergence of the iterative methods are determined . The methods are illustrated by a number of selected examples. The chapter ends with a set of questions.

Keywords: Fix point iterations, Newton's method, Secant method, Bisection method.

Introduction In this chapter, we shall solve the equation (*cf.* [3,22,24,25])

$$F(x) = 0, \quad a \leq x \leq b, \qquad (7.1)$$

by the following iterative methods:

- Fixed point iterations,
- Newton's method,

- Secant method,
- Bisection method.

The rate of convergence of these iterative methods is determined by the order which is defined as follows:

Definition 7.1 *Let $\{x_n\}$, $n = 0, 1, \ldots$; be a sequence convergent to a root α of the equation $F(x) = 0$, $x \in [a, b]$. An iterative method is said to be of order p if there exists a constant $C \neq 0$ such that*

$$\lim_{n \to \infty} \frac{|\alpha - x_{n+1}|}{|\alpha - x_n|^p} = C.$$

To illustrate the definition, let us consider the following equation:

$$x = f(x), \quad a \leq x \leq b,$$

where f is a given differentiable function and its derivative $f'(x) \neq 0$ in the interval (a, b). Suppose that the sequence

$$x_{n+1} = f(x_n), \quad n = 0, 1, \ldots; \tag{7.2}$$

is convergent to the root $\alpha \in [a, b]$, for any choice of the starting point $x_0 \in [a, b]$.
Obviously, we have

$$\alpha - x_{n+1} = f(\alpha) - f(x_n) = f'(\xi_n)(\alpha - x_n)$$

for certain ξ_n between α and x_n .
Hence

$$\frac{\alpha - x_{n+1}}{\alpha - x_n} = f'(\xi_n)$$

and

$$\lim_{n \to \infty} \frac{|\alpha - x_{n+1}|}{|\alpha - x_n|} = f'(\alpha).$$

Therefore, iteration method (7.2) has of order $p = 1$, since $f'(x) \neq 0$ in the interval (a, b). Let us note that the order of the method is greater than one, if $f'(\alpha) = 0$.

7.1   Fixed Point Iterations

In order to describe *fixed point iterations,* we write equation (7.1) in the following form:

$$x = f(x), \qquad a \le x \le b, \tag{7.3}$$

Then, $f(x)$ is *called iteration function.* There are many ways to write equation (7.1) in the iterative form (7.3). We present this in the following example:

Example 7.1 *Consider the equation*

$$x^2 - x - 4 = 0, \qquad -\infty < x < \infty. \tag{7.4}$$

Write this equation in the iterative form (7.3).

Obviously, equation (7.4) can be written in one of the following form:

$$x = x^2 - 4, \qquad f(x) = x^2 - 4, \qquad -\infty < x < \infty,$$

$$x = \sqrt{x + 4}, \qquad f(x) = \sqrt{x + 4}, \qquad x \ge -4,$$

$$x = \frac{4}{x - 1}, \qquad f(x) = \frac{4}{x - 1}, \qquad x \ne 1,$$

$$x = 1 + \frac{4}{x}, \qquad f(x) = 1 + \frac{4}{x}, \qquad x \ne 0.$$

We can approximate a root of equation (7.3) by the sequence of successive iterations

$$x_{n+1} = f(x_n), \qquad n = 0, 1, \ldots,$$

where x_0 is a starting value.
For example, the equation

$$4x - e^{-x} = 0, \qquad -\infty < x < \infty,$$

has the iterative form

$$x = \frac{1}{4} e^{-x}, \qquad -\infty < x < \infty.$$

Then, the sequence

$$x_{n+1} = \frac{1}{4}e^{-x_n}, \qquad n = 0, 1, ...,$$

converges to the root $\alpha = 0.2039...$ with an arbitrary starting value x_0.

Indeed, choosing $x_0 = 0$ as starting value, we calculate terms of the sequence

$$x_1 = 0,25$$
$$x_2 = 0.1947$$
$$x_3 = 0.2058$$
$$x_4 = 0.2035$$
$$x_5 = 0.2040$$
$$x_6 = 0.2039$$
$$x_7 = 0.2039$$
$$........$$

We can obtain the fixed point sequence by the following instructions:

```
f[x_]:=Exp[-x]/4;
N[NestList[f,0,7],4]
```

or

```
f[x_]:=Exp[-x]/4;
N[FixedPointList[f,0,7],4]
```

Let us observe that the above sequence converges to the root $\alpha = 0.2039...$ of the equation $x = \frac{1}{4}e^{-x}$, so that

$$\alpha = \frac{1}{4}e^{-\alpha}.$$

In general, if there exists an $\alpha \in [a, b]$ such that

$$\alpha = f(\alpha),$$

then α is called the *fixed point* of the function f.

The fixed point iteration sequence $x_{n+1} = f(x_n)$, $n = 0, 1, ...$, converges to α under assumptions specified in the following fixed point theorem:

Theorem 7.1 *Assume that the iteration function f satisfies the following conditions:*

1. *f is a continuous on the closed interval $[a, b]$ and continuously differentiable in the open interval (a, b),*

2. *f maps the interval $[a, b]$ into $[a, b]$, i.e., $f : [a, b] \rightarrow [a, b]$.*

3. *there exists a constant $K < 1$ such that*

$$\mid f'(x) \mid \leq K$$

for all $x \in (a, b)$,

Then the iteration function f has a unique fixed point $\alpha \in [a, b]$, and α is the limit of the fixed point sequence

$$x_{n+1} = f(x_n), \quad n = 0, 1, \ldots ;$$

The limit α is independent of the starting value x_0, so that

$$\lim_{n \to \infty} x_n = \alpha$$

for any $x_0 \in [a, b]$.

Proof. We shall present the proof of this theorem according to the following plan:

- First, we shall show that there exists a fixed point of f in the interval $[a, b]$,

- Secondly, we shall show that there is no more than one fixed point in the interval $[a, b]$,

- Thirdly, we shall show that the fixed point sequence

$$x_{n+1} = f(x_n), \quad n = 0, 1, \ldots ;$$

converges to the unique fixed point α.

Clearly, if $f(a) = a$ or $f(b) = b$, then f has a fixed point in $[a, b]$. Otherwise, let $f(a) \neq a$ and $f(b) \neq b$. By assumption 2, we note that $f(a) \in [a, b]$ and $f(b) \in [a, b]$. Therefore

$$f(a) > a \quad \text{and} \quad f(b) < b.$$

Then, the function

$$h(x) = f(x) - x, \quad a \le x \le b,$$

is positive at $x = a$ and negative at $x = b$, i.e.,

$$h(a) > 0 \quad \text{and} \quad h(b) < 0.$$

Since h is a continuous function and changes its sign in $[a, b]$, we conclude, by the intermediate-value theorem, that $h(x)$ must vanish somewhere in (a, b). Therefore, f has at least one fixed point in the interval $[a, b]$.

In order to prove that there is one fixed point of f in $[a, b]$, let us assume that there are two different fixed points α and β such that

$$\alpha = f(\alpha), \quad \beta = f(\beta), \quad \alpha \ne \beta. \tag{7.5}$$

Then, by the mean value theorem

$$f(\alpha) - f(\beta) = f'(\xi)(\alpha - \beta)$$

for certain $\xi \in (a, b)$.
By (7.5), we get

$$\alpha - \beta = f'(\xi)(\alpha - \beta).$$

Hence, the equality

$$f'(\xi) = 1.$$

contradicts assumption 3. Therefore, f has a unique fixed point in the interval $[a, b]$.

In order to prove that the fixed point iteration sequence $x_{n+1} = f(x_n)$, $n = 0, 1, \ldots$; is convergent to α, let us consider the error of n-th iteration

$$e_n = \alpha - x_n.$$

Then, we have

$$e_{n+1} = \alpha - x_{n+1} = f(\alpha) - f(x_n) = f'(\xi)(\alpha - x_n) = f'(\xi)e_n$$

for certain $\xi \in (a, b)$.

Hence, by assumption 3,

$$\mid e_{n+1} \mid \le K \mid e_n \mid, \quad n = 0, 1, \ldots;$$

As repeating application of this inequality, we obtain

$$\mid e_{n+1} \mid \le K \mid e_n \mid \le K^2 \mid e_{n-1} \mid \le \cdots \le K^{n+1} \mid e_0 \mid .$$

So that

$$\mid \alpha - x_n \mid \le K^n \mid \alpha - x_0 \mid . \tag{7.6}$$

Since $0 \le K < 1$, we get

$$\lim_{n \to \infty} K^n = 0$$

Hence, by (7.6)

$$\lim_{n \to \infty} x_n = \alpha.$$

This ends the proof.

In order to terminate the fixed point iteration sequence $\{x_n\}$, $n = 0, 1, \ldots$, we can use the following condition:

$$\mid x_{m+1} - x_m \mid \le \epsilon,$$

where m is the least integer for which the above condition hold, and ϵ is a given accuracy of the root α. Let as note that, the fixed point iterative method has order of convergence $p = 1$, if $f'(x) \ne 0$ for $x \in (a, b)$, (see (7.2)).

Example 7.2 *Consider the following equation:*

$$x = \frac{1}{4 + ax}, \quad x \in [0, 1] \tag{7.7}$$

(a). *Show that the iteration function satisfies the assumptions of the fixed point theorem for each $0 \le a < 16$.*

(b). *Compute the root of equation (7.7) for $a = 2$ with accuracy $\epsilon = 0.000001$.*

Solution (a) The iteration function

$$f(x) = \frac{1}{4 + ax}$$

is continuous for all $x \in [0, 1]$ and $a \in [0, 16)$, since then $4 + ax \geq 4$. Thus, assumption 1 is satisfied. For assumption 2, we have

$$0 \leq \frac{1}{4 + ax} \leq \frac{1}{4}, \quad \text{when} \quad 0 \leq x \leq 1, \quad \text{and} \quad 0 \leq a < 16.$$

Hence $f : [0, 1] \rightarrow [0, 1]$ and assumption 2 holds, too. Finally, assumption 3 is also satisfied, since we have

$$\mid f'(x) \mid = \frac{a}{(4 + ax)^2} \leq \frac{a}{16} < 1$$

for $0 \leq x \leq 1$ and $0 \leq a < 16$.

Thus, f satisfies all the assumptions of the fixed point theorem.

(b). Using the formula

$$x_{n+1} = \frac{1}{4 + 2x_n}, \quad n = 0, 1, \ldots$$

with $x_0 = 1$, we compute

$$x_1 = \frac{1}{4 + 2x_0} = 0.1666667$$

$$x_2 = \frac{1}{4 + 2x_1} = 0.2307692$$

$$x_3 = \frac{1}{4 + 2x_2} = 0.2241379$$

$$x_4 = \frac{1}{4 + 2x_3} = 0.2248062$$

$$x_5 = \frac{1}{4 + 2x_4} = 0.2247387$$

$$x_6 = \frac{1}{4 + 2x_5} = 0.2247455$$

$$x_7 = \frac{1}{4 + 2x_6} = 0.2247448$$

Clearly, the error

$$|x_{m+1} - x_m| = 0.0000007 < \epsilon = 0.000001 \quad \text{for} \quad m = 6.$$

Using `Mathematica`, we can find the root by the instruction

```
f[x_]:=1/(4+2*x);
N[FixedPoint[f,1,SameTest->(Abs[#1-#2]<0.000001&)]]
```

7.2   Newton's Method

We shall us consider the equation

$$F(x) = 0, \quad a \le x \le b, \tag{7.8}$$

under the following assumptions:

1. $F(x)$ is twice continuously differentiable function in the interval $[a, b]$,

2. the values $F(a)$ and F(b) have different signs, i.e.,

$$F(a)F(b) < 0$$

3. $F'(x)$ and $F''(x)$ do not change their signs in the interval $[a, b]$, that is, $F'(x)$ and $F''(x)$ are either positive or negative for all $x \in [a, b]$.

The equation (7.8) has a unique root $\alpha \in [a, b]$ if the conditions 1,2 and 3 are satisfied.

Indeed, by the intermediate value theorem, $F(x) = 0$ at certain $x = \alpha \in (a, b)$, since $F(x)$ is a continuous function in the interval $[a, b]$, and $F(a)F(b) < 0$. By assumption 3, we conclude that $F(x)$ is either decreasing or increasing in $[a, b]$. Thus, α must be a unique root of equation (7.8).

Derivation of Newton's method. Let $x_0 \in [a, b]$ be an arbitrarily chosen starting value. The equation of the line tangent at x_0 to the curve $y = F(x)$ is

$$y - F(x_0) = F'(x_0)(x - x_0).$$

As the first approximation of the root α, we consider the point of intersection of the line tangent with the x-axis. Then, for $y = 0$, we have

$$-F(x_0) = F'(x_0)(x - x_0).$$

and the first approximate value of the root α is

$$x_1 = x_0 - \frac{F(x_0)}{F'(x_0)}.$$

To evaluate x_2, as the next approximation of the root α, we find the point of intersection of the line tangent at x_1 to the curve $y = F(x)$ with the x-axis. Then, we have

$$y - F(x_1) = F'(x_1)(x - x_1).$$

Hence, for $y = 0$, we get

$$-F(x_1) = F'(x_1)(x - x_1),$$

and the next approximate value of the root α is

$$x_2 = x_1 - \frac{F(x_1)}{F'(x_1)}.$$

In a repeating process, we obtain the following Newton formula:

$$x_{n+1} = x_n - \frac{F(x_n)}{F'(x_n)}, \quad n = 0, 1, \ldots, m;$$

where $x_0 \in [a, b]$ is an arbitrarily chosen starting value, and m is the least integer for which [1]

$$\mid x_{m+1} - x_m \mid \leq \epsilon.$$

Analysis of convergence. We shall investigate convergence of Newton's method under the following conditions 1, 2 and 3.

As a consequence of these three conditions there are four the following cases:

Case I : $F'(x) > 0$ and $F''(x) > 0$, $x \in [a, b]$. In this case, $F(x)$ is an increasing and convex function.

---

[1] ϵ is a given accuracy to calculate α

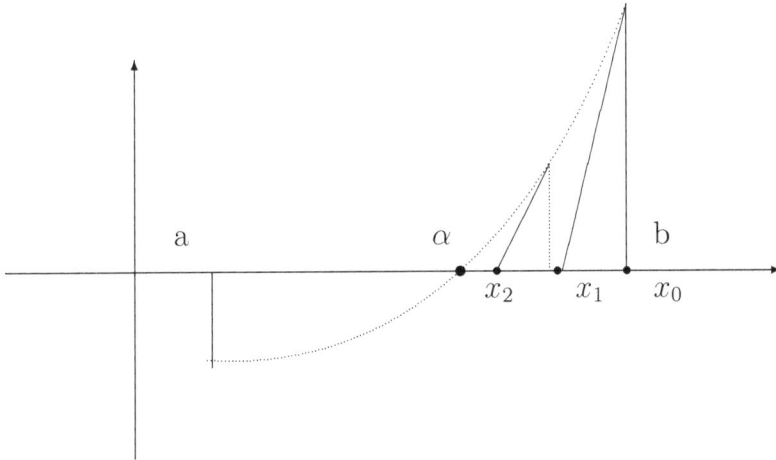

Fig. 7.1.

Case II : $F'(x) < 0$ and $F''(x) > 0$, $\quad x \in [a, b]$. In this, case $F(x)$ is a decreasing and convex function.

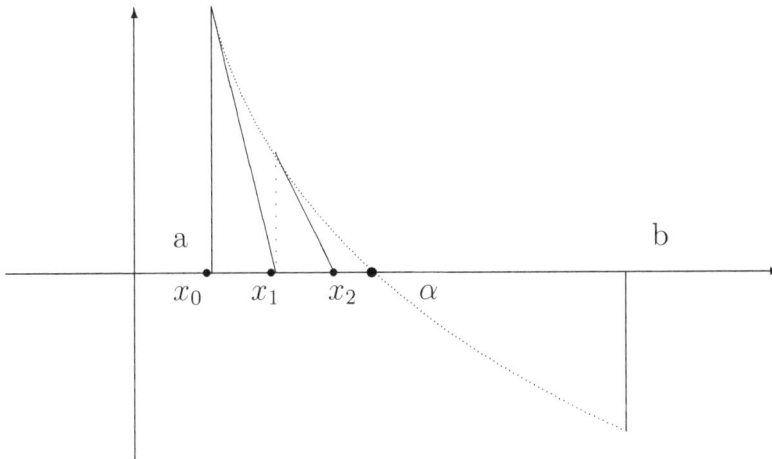

Fig. 7.2.

Case III : $F'(x) > 0$ and $F''(x) < 0$, $\quad x \in [a, b]$. In this case, $F(x)$ is an increasing and concave function.

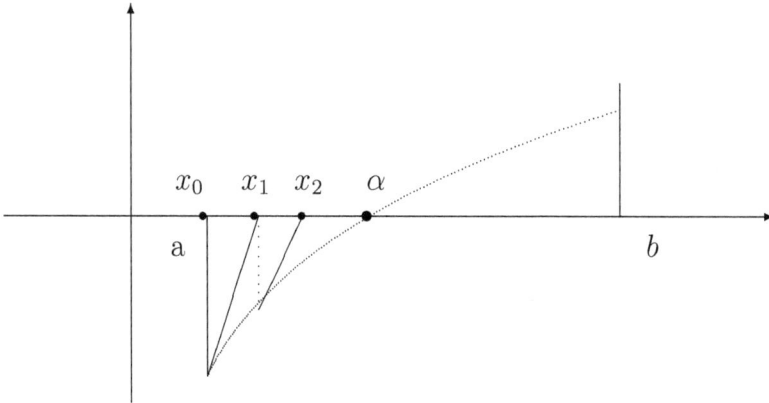

Fig. 7.3.

Case IV : $F'(x) < 0$ and $F''(x) < 0$, $x \in [a, b]$. In this case, $F(x)$ is a decreasing and concave function.

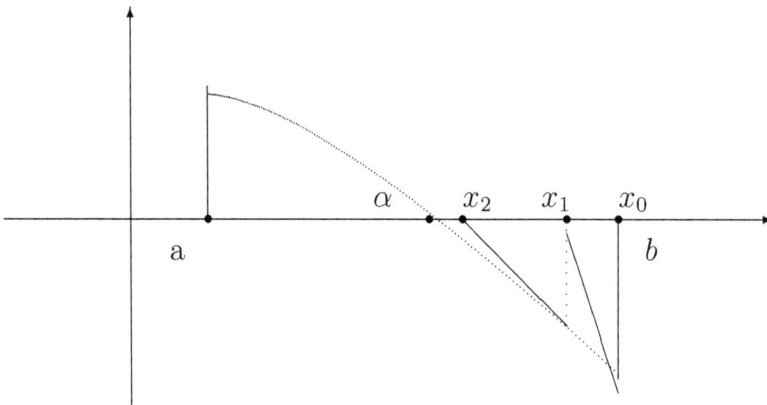

Fig. 7.4.

Obviously, we can choose the starting value x_0 in such a way that $F(x_0)F'(x_0) > 0$. Now, let us note that in cases I and IV when, either both $F'(x)$ and $F''(x)$ are positive or both

$F'(x)$ and $F''(x)$ are negative, the first approximation x_1 of root α is between α and x_0. Indeed, we have

$$x_1 = x_0 - \frac{F(x_0)}{F'(x_0)} < x_0.$$

Since

$$0 = F(\alpha) = F(x_0) + F'(x_0)(\alpha - x_0) + \frac{F''(\xi)}{2}(\alpha - x_0)^2 \quad (7.9)$$

for certain ξ between x_0 and α, we get

$$\alpha - x_1 = \alpha - x_0 + \frac{F(x_0)}{F'(x_0)} = -\frac{F''(\xi)}{2F'(x_0)}(\alpha - x_0)^2 < 0.$$

Therefore, we have the inequality

$$\alpha < x_1 < x_0.$$

In a similar way, we can show that successive Newton approximations satisfy the inequalities

$$\alpha < x_2 < x_1,$$
$$\alpha < x_3 < x_2,$$
$$\cdots \cdots \cdots \cdots$$
$$\alpha < x_{n+1} < x_n,$$
$$\cdots \cdots \cdots \cdots$$

From these inequalities, we note that

$$b \geq x_0 > x_1 > x_2 > \cdots > x_n > x_{n+1} > \cdots > \alpha.$$

This means that the sequence $\{x_n\}$, $n = 0, 1, \ldots$; is decreasing and lower bounded. Therefore, x_n converges to a limit $g \geq \alpha$, i.e., $x_n \to g$ when $n \to \infty$. In fact, $g = \alpha$, since

$$F(g) = \lim_{n \to \infty} F(x_n) = \lim_{n \to \infty} (x_n - x_{n+1})F'(x_n) = (g - g)F'(g) = 0$$

and α is a unique root of equation (7.8).

Now, let us choose x_0 such that $F(x_0)F'(x_0) < 0$ (for instance $x_0 = a$). In the cases II or III, when either $F'(x) < 0$

and $F''(x) > 0$ or $F'(x) > 0$ and $F''(x) < 0$, the first approximation x_1 lies between a and x_0. Indeed, we have

$$x_1 = x_0 - \frac{F(x_0)}{F'(x_0)} > x_0.$$

and

$$\alpha - x_1 = \alpha - x_0 + \frac{F(x_0)}{F'(x_0)} = -\frac{F''(\xi)}{2F'(x_0)} > 0,$$

for certain $\xi \in (a, b)$.
Hence

$$a \geq x_0 < x_1 < \alpha.$$

Likewise, we can show, that the next approximations lie also between x_0 and α and consequently, we note that

$$a \leq x_0 < x_1 < \cdots < x_n < x_{n+1} < \cdots < \alpha.$$

Thus, in cases II and III, the sequence $\{x_n\}$, $n = 0, 1, \ldots$; is increasing and upper bounded. Therefore, $x_n \to \alpha$ when $n \to \infty$.
Let us note that the iteration function of Newton's method is

$$f(x) = x - \frac{F(x)}{F'(x)},$$

so that, the following equations

$$F(x) = 0 \qquad \text{and} \qquad x = f(x), \qquad a \leq x \leq b,$$

are equivalent. Clearly the first derivative

$$f'(\alpha) = \frac{F(\alpha)F''(\alpha)}{[F'(\alpha)]^2} = 0.$$

This indicates the order of Newton's method $p \geq 2$. Indeed, by formula (7.9)), we have

$$\alpha - x_{n+1} = \alpha - x_n + \frac{F(x_n)}{F'(x_n)} = -\frac{F''(\xi)}{2F'(x_n)}(\alpha - x_n)^2.$$

Then

$$\frac{\alpha - x_{n+1}}{(\alpha - x_n)^2} = \frac{F''(\xi_n)}{2F'(x_n)} \tag{7.10}$$

for certain ξ_n between α and x_n.

Hence

$$\lim_{n \to \infty} \frac{|\alpha - x_{n+1}|}{|\alpha - x_n|^2} = \lim_{n \to \infty} \frac{|F''(\xi_n)|}{2|F'(x_n)|} = \frac{F''(\alpha)|}{2|F'(\alpha)|} \neq 0.$$

Therefore, the order of Newton's method $p = 2$ since

$$F''(\alpha) \neq 0$$

.

Now, we can state the following theorem:

Theorem 7.2 *If the conditions 1, 2 and 3 are satisfied then Newton's method is convergent with order $p = 2$ to a unique root α for any choice of the starting point $x_0 \in [a, b]$.*

Example 7.3 *Write Newton's algorithm to evaluate square root $\sqrt{c}$ for any $c > 0$.*

Let $F(x) = x^2 - c$ for $x > 0$. Then, we can find the square root $\sqrt{c}$ solving the equation

$$x^2 - c = 0, \quad x > 0, \quad c > 0.$$

Obviously, $F(a) = -c < 0$, for $a = 0$ and $F(b) = b^2 - c > 0$, for $b > c$, that is $F(a)F(b) < 0$. Also, we have $F'(x) = 2x > 0$, $x > 0$, and $F''(x) = 2 > 0$. Thus, the function $F(x)$ satisfies conditions 1, 2 and 3 when $x > 0$. Therefore, by the theorem, Newton sequence

$$x_{n+1} = x_n - \frac{x_n^2 - c}{2x_n} = \frac{1}{2}(x_n + \frac{c}{x_n}), \quad n = 0, 1, \dots;$$

converges quadratically to the root $\sqrt{c} = \alpha$, when $n \to \infty$ for any starting value $x_0 > 0$.

Applying the above algorithm to find $\sqrt{3}$ with the accuracy of four digits after decimal point, we obtain

$$x_0 = 1$$
$$x_1 = \frac{1}{2}(x_0 + \frac{3}{x_0}) = 2$$
$$x_2 = \frac{1}{2}(x_1 + \frac{3}{x_1}) = 1.75$$
$$x_3 = \frac{1}{2}(x_2 + \frac{3}{x_2}) = 1.7321$$
$$x_4 = \frac{1}{2}(x_3 + \frac{3}{x_3}) = 1.73205$$
$$x_5 = \frac{1}{2}(x_4 + \frac{3}{x_4}) = 1.73205$$

We can repeat the above calculation in `Mathematica` by executing the following recursive formula

```
sol[1]=1.;
sol[n_]:=sol[n]=(sol[n-1]+3/sol[n-1])/2;
Table[sol[n],{n,1,5}];
```

7.3   Secant Method

The secant method is a simple modification of Newton's method. Namely, let us replace first derivative $F'(x_n)$ which appears in Newton's method by the divided difference

$$F'(x_n) \approx \frac{F(x_n) - F(x_{n-1})}{x_n - x_{n-1}}.$$

Then, we obtain the secant method

$$x_{n+1} = x_n - \frac{F(x_n)(x_n - x_{n-1})}{F(x_n) - F(x_{n-1})} \qquad (7.11)$$

and its equivalent form

$$x_{n+1} = \frac{x_{n-1}F(x_n) - x_n F(x_{n-1})}{F(x_n) - F(x_{n-1})}, \qquad (7.12)$$

for $n = 1, 2, \ldots$;

In order to start with the secant method, we have to choose two initial values x_0 and x_1. Starting with these initial values, we can produce the sequence $\{x_n\}$, $n = 1, 2, \ldots$; using formula (7.12). Let us note that to evaluate x_{n+1}, we can use the already known value $F(x_{n-1})$ and the new value $F(x_n)$ must be calculated. Thus, at each iteration $F(x)$ is evaluated one time. But, using Newton's method both values $F(x_n)$ and $F'(x_n)$ must be evaluated at each iteration. If $F'(x)$ is expensive in terms of arithmetic operations, then Newton's method may occur to be slower than the secant method when a computer is used, in spite of its higher order of convergence.

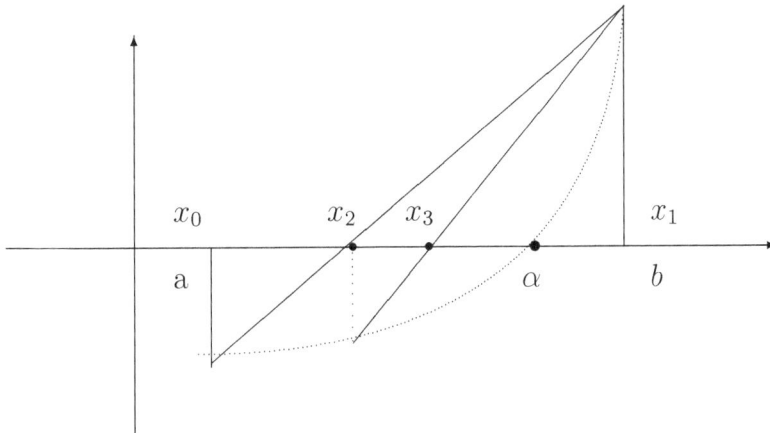

Fig. 7.5

Convergence of the secant method. We assume that conditions 1, 2 and 3 are satisfied. Then, there is one root α of equation $F(x) = 0$ in the interval $[a, b]$, and there are four possible cases I, II, III and IV (see figures **7.1, 7.2, 7.3, 7.4.**) First, let us consider cases I and IV, when either $F'(x) > 0$ and $F''(x) > 0$ or $F'(x) < 0$ and $F''(x) < 0$ in the interval

$[a, b]$. So, we can choose $x_0 < x_1$ such that $F(x_0) < 0$ and $F(x_1) > 0$ in case I or $F(x_0) > 0$ and $F(x_1) < 0$ in case IV, (for instance $x_0 = a$ and $x_1 = b$).

Now, we shall show that the sequence $\{x_n\}$, $n = 2, 3, \ldots$; produced by the secant method (see (7.11) and (**Fig. 7.5**), converges to the root α, i.e., $x_n \to \alpha$, when $n \to \infty$. In cases I and IV, the following inequalities hold:

$$x_0 < x_2 < x_3 < x_5 < \cdots < \alpha < \cdots < x_{10} < x_7 < x_4 < x_1$$
$$(7.13)$$

Thus, every third term x_{3k+1}, $k = 0, 1, \ldots$; of the sequence $\{x_n\}$, $n = 0, 1, \ldots$; lies on the right hand side of the root α. We shall prove that

$$x_0 < x_2 < x_3 < \alpha < x_4 < x_1. \qquad (7.14)$$

We note that

$$x_2 = x_1 - \frac{F(x_1)(x_1 - x_0)}{F(x_1) - F(x_0)} = x_1 - \frac{F(x_1)}{F'(\xi_0)} < x_1$$

and

$$\begin{aligned}
x_2 &= x_0 + (x_1 - x_0) - \frac{F(x_1)(x_1 - x_0)}{F(x_1) - F(x_0)} \\
&= x_0 - \frac{F(x_0)(x_1 - x_0)}{F(x_1) - F(x_0)} = x_0 - \frac{F(x_0)}{F'(\xi_0)} > x_0,
\end{aligned}$$

for certain $\xi_0 \in (x_0, x_1)$.

Hence

$$x_0 < x_2 < x_1.$$

By the assumption, $F(x)$ is a convex (concave) function in the interval $[a, b]$. Therefore

$$F(x) < F(x_0) + \frac{F(x_1) - F(x_0)}{x_1 - x_0}(x - x_0), \quad \text{when} \quad F''(x) > 0,$$

or

$$F(x) > F(x_0) + \frac{F(x_1) - F(x_0)}{x_1 - x_0}(x - x_0), \quad \text{when} \quad F''(x) < 0$$

for $x_0 < x < x_1$.
Let $x = x_2$. Then

$$F(x_2) < F(x_0) + \frac{F(x_1) - F(x_0)}{x_1 - x_0}(x_2 - x_0) = 0,$$

or

$$F(x_2) > F(x_0) + \frac{F(x_1) - F(x_0)}{x_1 - x_0}(x_2 - x_0) = 0.$$

This means that $x_2 < \alpha$ in cases I and IV, so that

$$x_0 < x_2 < \alpha. \tag{7.15}$$

Now, we shall show that

$$x_2 < x_3 < \alpha.$$

Indeed, we have

$$x_3 = x_2 - \frac{F(x_2)(x_2 - x_1)}{F(x_2) - F(x_1)} = x_2 - \frac{F(x_2)}{F'(\xi_1)} > x_2$$

and

$$x_3 = x_1 + (x_2 - x_1) - \frac{F(x_2)(x_2 - x_1)}{F(x_2) - F(x_1)} = x_1 - \frac{F(x_1)}{F'(\xi_1)} < x_1$$

for certain $\xi_1 \in (x_2, x_1)$.
Hence

$$x_2 < x_3 < x_1. \tag{7.16}$$

Again, in case I

$$F(x) < F(x_1) + \frac{F(x_2) - F(x_1)}{x_2 - x_1}(x - x_1)$$

and in case IV

$$F(x) > F(x_1) + \frac{F(x_2) - F(x_1)}{x_2 - x_1}(x - x_1)$$

for $x_2 < x < x_1$.
Substituting $x = x_3$ to these inequalities, we obtain

$$F(x_3) < F(x_1) + \frac{F(x_2) - F(x_1)}{x_2 - x_1}(x_3 - x_1) = 0.$$

or

$$F(x_3) > F(x_1) + \frac{F(x_2) - F(x_1)}{x_2 - x_1}(x_3 - x_1) = 0.$$

Hence, $x_3 < \alpha$. By (7.15) and (7.16), we have

$$x_0 < x_2 < x_3 < \alpha < x_1.$$

To complete the proof of inequality (7.14), we shall show that

$$\alpha < x_4 < x_1.$$

Let us note that

$$x_4 = x_3 - \frac{F(x_3)(x_3 - x_2)}{F(x_3) - F(x_2)} = x_3 - \frac{F(x_3)}{F'(\xi_1)} > x_3$$

and

$$F(x) > F(x_2) + \frac{F(x_3) - F(x_2)}{x_3 - x_2}(x - x_2)$$

for $x < x_2$ or $x > x_3$.

Hence, for $x = x_4 > x_3$, we have

$$F(x_4) > F(x_2) + \frac{F(x_3) - F(x_2)}{x_3 - x_2}(x_4 - x_2) = 0.$$

Since $F(x_4) > 0$, $\alpha < x_4$. Clearly, $x_4 < x_1$. This completes the proof of inequality (7.14).

Replacing x_0 by x_3 and x_1 by x_4, we can repeat the above estimates to get the following inequality:

$$x_3 < x_5 < x_6 < \alpha < x_7 < x_4. \tag{7.17}$$

Next, we may replace x_3 by x_6 and x_4 by x_7, to obtain

$$x_6 < x_8 < x_9 < \alpha < x_{10} < x_7 \tag{7.18}$$

Finally, by the repetition of the replacement x_{3k-3} by x_{3k} and x_{3k-2} by x_{3k+1} for $k = 1, 2, \ldots;$, we can arrive at the inequalities

$$x_{3k} < x_{3k+2} < x_{3k+3} < \alpha < x_{3k+4} < x_{3k+1}, \quad k = 1, 2, \ldots; \tag{7.19}$$

Hence, we obtain the inequality (7.13).

Then, in the case I and IV, the sequence $\{x_n\},\ \ n = 2, 3, \ldots;$ produced by the secant method is splited into two monotonic and bounded subsequences $\{x_n\}$, when $n \ mod \ 3 \neq 0$, and $\{x_n\}$, when $n \ mod \ 3 = 0,\ \ n = 0, 1, \ldots;$ Therefore, both subsequences are convergent. To complete the analysis of convergence of the second method, we should show that these subsequences are convergent to the same limit α. Namely, let

$$\lim_{n \to \infty,\ n \neq 0 mod 3} x_n = g.$$

From the formula

$$x_{n+1} = x_n - \frac{F(x_n)(x_n - x_{n-1})}{F(x_n) - F(x_{n-1})} = x_n - \frac{F(x_n)}{F'(\xi_n)}, \quad \xi_n \in (x_{n-1}, x_n),$$

it follows that

$$\lim x_{n+1} = \lim x_n - \lim \frac{F(x_n)}{F'(\xi_n)}.$$

Hence

$$g = g - \frac{F(g)}{F'(g)} \quad \text{and} \quad F(g) = 0.$$

Since the equation $F(x) = 0$ has unique root $\alpha \in [a, b]$, we get $g = \alpha$. In the same way, we can show that

$$\lim_{n \to \infty,\ n \ mod \ 3 = 0} x_n = \alpha$$

The proof of convergence of the sequence $\{x_n\},\ \ n = 2, 3, \ldots;$ produced by the secant method, in cases II and III, is dealt within questions.

Order of convergence of the secant method. Let us note that the equation of the secant line through the points

$$(x_{n-1}, F(x_{n-1})) \quad \text{and} \quad (x_n, F(x_n))$$

is the interpolating polynomial

$$P_1(x) = \frac{x - x_n}{x_{n-1} - x_n} F(x_{n-1}) + \frac{x - x_{n-1}}{x_n - x_{n-1}} F(x_n).$$

Therefore

$$F(x) = P_1(x) + \frac{F''(\xi_n)}{2!}(x - x_{n-1})(x - x_n), \quad \xi_n \in (x_{n-1}, x_n).$$

Hence

$$0 = F(\alpha) = P_1(\alpha) + \frac{F''(\xi_n)}{2!}(\alpha - x_{n-1})(\alpha - x_n).$$

Thus, we have

$$P_1(\alpha) = -\frac{F''(\xi_n)}{2!}(\alpha - x_{n-1})(\alpha - x_n). \tag{7.20}$$

On the other hand

$$\begin{aligned}
P_1(\alpha) &= \frac{\alpha - x_n}{x_{n-1} - x_n}F(x_{n-1}) + \frac{\alpha - x_{n-1}}{x_n - x_{n-1}}F(x_n) \\
&= \frac{x_n F(x_{n-1}) - x_{n-1}F(x_n)}{x_n - x_{n-1}} + \alpha\frac{F(x_n) - F(x_{n-1})}{x_n - x_{n-1}}.
\end{aligned}$$

By the mean value theorem

$$\frac{F(x_n) - F(x_{n-1})}{x_n - x_{n-1}} = F'(\eta_n),$$

for certain $\eta_n \in (x_{n-1}, x_n)$. From formula (7.12), we get

$$\frac{x_n F(x_{n-1}) - x_{n-1}F(x_n)}{x_n - x_{n-1}} = -x_{n+1}F'(\eta_n).$$

Hence

$$P_1(\alpha) = -x_{n+1}F'(\eta_n) + \alpha F'(\eta_n) = (\alpha - x_{n+1})F'(\eta_n). \tag{7.21}$$

Combining (7.20) and (7.21), we obtain the following relation between the errors $e_{n+1} = \alpha - x_{n+1}$, $e_{n-1} = \alpha - x_{n-1}$ and $e_n = \alpha - x_n$:

$$e_{n+1} = \frac{F''(\xi_n)}{2F'(\eta_n)}e_{n-1}e_n$$

for $n = 0, 1, \ldots$;

Now, we note the following:

$$\frac{\mid e_{n+1} \mid}{\mid e_n \mid^p} = \frac{\mid F''(\xi_n) \mid}{2 \mid F'(\eta_n) \mid} \mid e_n \mid^{1-p} \mid e_{n-1} \mid = \frac{\mid F''(\xi_n) \mid}{2 \mid F'(\eta_n) \mid} [\frac{\mid e_n \mid}{\mid e_{n-1} \mid^p}]^{\beta},$$

for $\beta = 1 - p$ and $\beta p = -1$.

Hence, we have

$$(1 - p)p = -1 \quad \text{and} \quad p^2 - p - 1 = 0.$$

The positive root of the above equation is : $p = \frac{1+\sqrt{5}}{2}$.
Choosing $p = \frac{1+\sqrt{5}}{2}$ and $\beta = \frac{2}{1+\sqrt{5}}$, we obtain the following equation:

$$y_{n+1} = a_n y_n,$$

where

$$a_n = \frac{\mid F''(\xi_n) \mid}{2 \mid F'(\eta_n) \mid}, \quad y_{n+1} = \frac{\mid e_{n+1} \mid}{\mid e_n \mid^p}, \quad y_n = \frac{\mid e_n \mid}{\mid e_{n-1} \mid^p}.$$

Then, $y_n \to \gamma$, when $n \to \infty$, where γ is the fixed point of the equation

$$x = ax^{-\frac{1}{p}}, \quad a = \lim a_n.$$

Let us note that

$$\gamma = a^{\frac{1}{p}} = [\frac{\mid F''(\alpha) \mid}{2 \mid F'(\alpha) \mid}]^{\frac{1}{p}}.$$

This shows that the error e_n satisfies the following:

$$\lim_{n \to \infty} \frac{\mid e_{n+1} \mid}{\mid e_n \mid^p} = [\frac{\mid F''(\alpha) \mid}{2 \mid F'(\alpha) \mid}]^{\frac{1}{p}} = constant \neq 0,$$

for $p = \dfrac{1 + \sqrt{5}}{2}$. The secant method has order

$$p = \frac{1 + \sqrt{5}}{2} \approx 1.618.$$

In order to stop the iterations, we can use the same condition as in Newton's method. Namely, we stop the secant iterations at the least m such that

$$\mid x_{m+1} - x_m \mid \leq \epsilon.$$

Example 7.4 *Solve the equation*

$$e^{-x} - 2x = 0, \qquad 0 \le x \le 1,$$

by the secant method with accuracy $\epsilon = 0.01$.

Solution. The function $F(x) = e^{-x} - 2x, \quad x \in [0, 1]$ satisfies assumptions 1, 2 and 3. Indeed, $F(x)$ is infinitely times differentiable on the real line,. $F(0)F(1) = 1 * (e^{-1} - 2 < 0$ and $F'(x) = -e^{-x} - 2 < 0, \quad F''(x) = e^{-x} > 0$ for all $x \in [0, 1]$. In this example, the algorithm of the secant method is

$$x_{n+1} = \frac{x_{n-1}(exp(-x_n) - x_n) - x_n(exp(-x_{n-1}) - x_{n-1})}{exp(-x_n) - 2x_n - exp(-x_{n-1}) - 2x_{n-1}}$$

for $n = 1, 2, \ldots$; where x_0 and x_1 are starting values. Let the staring values $x_0 = 0$ and $x_1 = 1$. Then, we have

$$x_2 = \frac{0 * exp(-1) - 1}{exp(-1) - 3} = 0.3799218$$

$$x_3 = \frac{exp(-0.3799218) - 0.3799218 * exp(-1)}{exp(-0.3799218) - exp(-1) - 2(1 - 0.3799218)} = 0.349667$$

$$x_4 = \frac{0.3799218 * exp(-0.349667) - 0.349667 * exp(-0.3799218)}{exp(-0.349667) - exp(-0.3799218) + 2(0.3799218 - 0.349667)}$$

$$= 0.3517$$

We reach the accurate at $m = 3$, since then

$$|x_4 - x_3| = 0.3517 - 0.3497 = 0.002.$$

Then, we consider $\alpha \approx 0.3517$ as the approximate value of the root.

Also, we can find the approximate value of the root α by the following **Mathematica** program

```
f[x_]:=Exp[-x]-2*x;
sol[1]=0; sol[2]=1;
sol[n_]:=sol[n]=
N[(sol[n-2]*f[sol[n-1]]-sol[n-1]*f[sol[n-2]])/
(f[sol[n-1]]-f[sol[n-2]])];
Table[sol[n],{n,3,6}]
```

7.4   Bisection Method

Let $F(x)$ be a continuous function in the interval $[a, b]$ such that $F(a)F(b) < 0$. This assumption guarantees the existence of at least one root in the interval $[a, b]$. The idea of bisection method is based on the partition of the interval $[a, b]$ into two equal subintervals. Then, we choose the subinterval, where the root is allocated. The chosen subinterval is divided into two equal subintervals again, and the new subinterval with the root α is to be divided. We continue this partition of the interval $[a, b]$ as long as the chosen subinterval has the length smaller than a given accuracy ϵ and it contains the root.

Algorithm of the bisection method

Set $a_0 = a$, $b_0 = b$, $n = 0$
repeat

$$d_n = \frac{a_n + b_n}{2}$$

$$\text{if} \ \ F(a_n)F(d_n) < 0$$

$$\text{then set} \ \ a_{n+1} = a_n, \ \ b_{n+1} = d_n$$

$$\text{else set} \ \ a_{n+1} = d_n, \ \ b_{n+1} = b_n,$$

$$n := n + 1,$$

until

$$\frac{b_{n+1} - a_{n+1}}{2} = \frac{b - a}{2^{n+1}} \le \epsilon.$$

Then, the approximate root is

$$\alpha \approx \frac{b_{n+1} + a_{n+1}}{2}.$$

The bisection method has order $p = 1$.

Example 7.5 *Find root of the equation*
$$e^{-x} - 2x = 0, \quad 0 \le x \le 1,$$
by the bisection method with accuracy $\epsilon = 0.02$.

Solution. We note that $F(x) = e^{-x} - 2x$ is a continuous function and $F(0)F(1) = 1 * (e^{-1} - 2) < 0$. Following the algorithm, we compute

$$a_0 = 0, \quad b_0 = 1, \quad \epsilon = 0.02$$

for $n = 0$, $\quad d_0 = \dfrac{a_0 + b_0}{2} = \dfrac{0 + 1}{2} = 0.5,$

$F(a_0)F(d_0) = -0.39, \quad a_1 = 0, \quad b_1 = 0.5$

for $n = 1$, $\quad d_1 = \dfrac{a_1 + b_1}{2} = \dfrac{0 + 0.5}{2} = 0.25$

$F(a_1)F(d_1) = 0.078, \quad a_2 = 0.25, \quad b_2 = 0.5$

for $n = 2$, $\quad d_2 = \dfrac{a_2 + b_2}{2} = \dfrac{0.25 + 0.5}{2} = 0.375$

$F(a_2)F(d_2) = -0.017 \quad a_3 = 0.25, \quad b_3 = 0.375$

for $n = 3$, $\quad d_3 = \dfrac{a_3 + b_3}{2} = \dfrac{0.25 + 0.375}{2} = 0.3125$

$F(a_3)F(d_3) = 0.011, \quad a_4 = 0.3125, \quad b_4 = 0.375$

for $n = 4$, $\quad d_4 = \dfrac{a_4 + b_4}{2} = \dfrac{0.3125 + 0.375}{2} = 0.34375$

$F(a_4)F(d_4) = 0.0005, \quad a_5 = 0.34375, \quad b_5 = 0.375$

We reach the accuracy at $n = 5$, since
$$\frac{b_5 - a_5}{2} = \frac{0.375 - 0.34375}{2} = 0.015625,$$
Then, we get the approximate value of the root

$$\alpha \approx \frac{0.34375 + 0.375}{2} = 0.359375 \; .$$

We can also find the approximate value of the root α by bi-section method using the following in **Mathematica** program

```
f[x_]:=Exp[-x]-2*x;
a=0.; b=1.;
n=5; Do[{d=(a+b)/2,Print[d],
If[f[a]*f[b]<0,b=d,a=d]},{n}];
Print["The approximate value of the root is:
                              ",(a+b)/2];
```

7.5 Exercises

Question 7.1 .

1. (a) *State the fixed point theorem.*

   (b) *Show that the iteration function of the equation*

   $$x = \frac{a}{6 + ax}, \quad 0 \le x \le 1,$$

   satisfies the assumptions of the fixed point theorem for $0 \le a < 6$.

   (c) *Compute the root of the equation with accuracy $\epsilon = 0.01$ and $a = 3$ using the fixed point iterations.*

Question 7.2 .

1. (a) *Show that the equation*

   $$x = \frac{1}{4 + x^2 + a^2}, \quad 0 \le x \le 1, \quad\quad (7.22)$$

   possesses a unique root in the interval $[0, 1]$ for each $a \ge 0$.

   (b) *Show that the sequence*

   $$x_{n+1} = \frac{1}{4 + x_n^2 + a^2}, \quad n = 0, 1, \dots ;$$

   is convergent to the root of the equation (7.22) starting with any value $x_0 \in [0, 1]$ and evaluate the root with accuracy $\epsilon = 0.01$, when $a = 1$.

Question 7.3 .

1. (a) *For which values of the parameter β the iteration function of the equation*

   $$x = \frac{3}{3 + \beta^2\, x}, \quad\quad x \in [0, 1],$$

   satisfies the assumptions of the fixed point theorem.

(b) *Evaluate the approximate solution of the equation*

$$x = \frac{3}{3+x}, \qquad x \in [0, 1],$$

with accuracy $\epsilon = 0.005$ using minimum number of iterations.

Question 7.4 .

Consider the equation

$$e^{-x} - 8x = 0, \qquad -\infty < x < \infty.$$

1. (a) *Show the sequence determined by Newton's method for this equation is monotone for any choice of the starting value x_0.*

   (b) *Solve the equation by Newton's method with accuracy $\epsilon = 0.0001$.*

Question 7.5 .

1. (a) *Show that the Newton's method is convergent when it applied to the equation*

   $$ln(1 + x) = 0.5, \qquad 0 \leq x \leq 1.$$

   (b) *Write down an algorithm for Newton's method to solve the equation.*

   (c) *Show that the error of iteration satisfies the inequality*

   $$|x_n - \alpha| \leq |x_0 - \alpha|^{2^n}.$$

   (d) *Using the algorithm, find the root α of the equation with accuracy $\epsilon = 0.001$.*

Question 7.6 .

1. (a) *Write down conditions under which the Newton's method is convergent.*

(b) Let
$$f(x) = x - \frac{F(x)}{F'(x)}, \qquad a \leq x \leq b$$

be the iteration function of the Newton's method, where $F(x)$ satisfies the conditions you have stated in (a).

Show that the following equations
$$F(x) = 0 \quad \text{and} \quad x = f(x), \qquad a \leq x \leq b,$$

are equivalent. What is the value of $f'(\alpha)$ if
$$F(\alpha) = 0 \ ?$$

(c) Consider the following equation:
$$2x - \sqrt{1 + x^2} = 0, \qquad 0 \leq x \leq 1,$$

Show that the function $F(x) = 2x - \sqrt{1 + x^2}$ satisfies the conditions of convergence of the Newton's method in the interval $[0, 1]$.

Solve this equation by the Newton's method with accuracy $\epsilon = 0.02$.

Question 7.7 .
Consider the following equation:
$$F(x) = 0, \quad a \leq x \leq b.$$

where $F(x)$ is twice continuously differentiable function in the interval $[a, b]$ and satisfies the conditions:

1. $F(a)F(b) < 0$,
2. $F'(x) < 0$ and $F''(x) > 0$ for $x \in [a, b]$.

Show that the secant method is convergent for any choice of the starting values x_0 and x_1.

Question 7.8 .
Solve the equation
$$e^{-2x} - 4x = 0, \qquad 0 \leq x \leq 1,$$

with accuracy $\epsilon = 0.005$, by

1. (a) *fixed point iterations*
   (b) *Newton's method*
   (c) *secant method*
   (d) *bisection method*

using minimum number of iterations.

References

[1] Ahlberg, J.H., Nilson, E.N., & Walsh, J.L., (1967), The Theory of Splines and Their Applications, Academic Press.

[2] Aikinson Kendell, (1997), An Introdution to Numerical Analysis, John Wiley and Sons Publishershing.

[3] Berezin, I.S. & Zidkov, N.P. (1962), Numerical Methods, v. I & II, Phys. Mat. Literature.

[4] Cheney, W. & Kincaid, D., (1994), Numerical Mathematics & Computing, Brooks Cole Publishing Company.

[5] Conte, S. D., & de Boor, C., (1983), Elementary Numerical Analysis, Algorithmic Approach, McGraw-Hill.

[6] Douglas Faires and Richard L. Burden, (2004), Numerical Methods, Brooks cole Publishing.

[7] Epperson James, (2001), An Introduction to Numerical Methods and Analysis, John Wiley and Sons Publishing.

[8] Gerald Curtis, E.and Wheatley P.O. (2003), Applied Numerical Analysis, Pearson Publishers.

[9] Hall, C.A. (1968), On error bounds for spline interpolation, J. Approx. Theory, 1.

[10] Haauseholder A.S. (2006), Principles of Numerical Analysis, Dver Publishers.

[11] Henrici, P. (1962), Discrete Variable Methods in Ordinary Differential Equations, John Wiley & Sons.

[12] Hohmann A. and Deuflhard P. (2010), Numerical Analysis in Modern Scietific Computing: An Introduction, Springer Publishers.

[13] Linz, P.(1994), Theoretical Numerical Analysis: An Introduction to Advanced Techniques, Dver Publications.

[14] Lukaszewicz, J. & Warmus, M., (1956), Numerical and Graphical Methods, Biblioteka Matematyczna,

[15] Moris, J.L., (1983), Computational Methods in Numerical Analysis, John Wiley & Sons.

[**16**] Neumair A. (2012), Introduction to Numerical Analysis, Cambride University Press.

[**17**] Ortega J. M. (1990), Numerical Analysis, A Second Course, SIAM 3, In Applied Mathematics.

[**18**] Phillips, G.M., & Taylor, P.J., (1973), Theory and Applications of Numerical Methods, Academic Press.

[**19**] Powell, M.J.D., (1981), Approximation Theory and Methods, Cambridge University Press.

[**20**] Phillips G.M. and Taylor P.J. (1996), Numerical Analysis,Academic Press.

[**21**] Prenter, P.M. (1975), Splines and Variational Methods, Wiley-Interscience Publications.

[**22**] Ralston, A., (1965), A First Course in Numerical Analysis, McGraw-Hill.

[**23**] Sauer T. (2005), Numerical Analysis, Pearson Publishers.

[**24**] Scott, L. (2011), Numerical Analysis, Princeton University Press.

[**25**] Stoer, J. & Bulrisch R. (1980), Introduction to Numerical Analysis, Springer-Verlag

[**26**] Watson, G.A., (1980), Approximation Theory and Numerical Analysis, John Wiley & Sons.

[**27**] Walter Gautschi, (2012), Numerical Analysis, Birkhause Publications.

[**28**] Wolfram, S., (1992), Mathematica a System for Doing Mathematics by Computers, Addison-Wesley Publishing Company.

[**29**] M. Zlamal, (1969), On some Finite Element Procedures for Solving Second order Boundary Value Problems, Numer. Math. 14(1969), pp. 42-48.

Index

www.ingramcontent.com/pod-product-compliance
Lightning Source LLC
Chambersburg PA
CBHW050830220326
41598CB00006B/341